MECHANICAL SEAL PRACTICE FOR IMPROVED PERFORMANCE

Members of Working Party and List of Contributors

Dr A. G. H. Coombs (Chairman) Sterling Fluid Products
R. C. Colvin (Secretary) The Institution of Mechanical Engineers
T. N. Cleaver Flexibox Limited
P. J. Dolan BP International Limited
T. A. Dziewulski Foster Wheeler Energy Limited
J. K. Frew Weir Pumps Limited
A. J. Milne National Engineering Laboratory
Dr B. S. Nau British Hydromechanics Research Association
J. M. Plumridge E G & G Sealol
R. F. Polwarth E G & G Sealol
P. R. Rogers Crane Packing Limited
Dr J. D. Summers-Smith Consultant
N. M. Wallace Flexibox Limited
B. J. Woodley St Regis Paper Company (UK) Limited (formerly Portals Limited)

Mechanical seal practice for improved performance

Edited by

J. D. SUMMERS-SMITH

IMechE Guides for the Process Industries

Published by
Mechanical Engineering Publications Limited for
The Institution of Mechanical Engineers
LONDON

The publishers are not responsible for any statement made in this publication. Data, discussion and any conclusions developed by authors are for information only and are not intended for use without independent substantiating investigation on the part of the potential users.

British Library Cataloguing in Publication Data

Mechanical seal practice for improved
performance
1. Engineering components: Mechanical
seals
I. Summers-Smith, J. D. II. Series
621.8′85

ISBN 0–85298–671–8

Typeset by Paston Press, Loddon, Norfolk

Printed in Great Britain at the Alden Press, Oxford

CONTENTS

PREFACE

Since their commercial development some 50 years ago, mechanical seals have been increasingly used for sealing rotating shafts. Today, they are the predominant type of seal found on centrifugal pumps, compressors, and similar machines. In terms of sealing efficiency and versatility of application, they represent a significant improvement over former sealing techniques, for example, gland packing, which they have now largely displaced. However, despite their widespread use, there has been a growing concern amongst users about their performance in service. Although mechanical seals are capable of operating efficiently over a period of years without the need for maintenance attention in process plant applications, and do so in many cases, the fact remains that the behaviour of seals in general is unpredictable, involving a significant incidence of unexplained 'infant mortality' and random failures. Currently, the mean time between failure of seals in process applications appears to be between eight and thirteen months, well below, for example, the lives expected from rolling contact bearings. This represents a significant penalty in terms of downtime and cost of maintenance, together with potential hazards, depending on the liquid being contained.

The Institution fo Mechanical Engineers, as part of its 'Technology for Industry' campaign, held seminars in May 1983 and again in March 1984 in an attempt to clarify the situation. These meetings were attended by seal manufacturers, pump manufacturers, and users, all of whom were invited to express their views. Discussions were wide ranging, confirming concern that a genuine problem does exist in the application of mechanical seals, particularly in the more demanding and more critical duties. The very considerable amount of research into seal behaviour was noted. The solution of the problem, however, was not considered to be related to any necessity for further research at this stage, but rather through the exploitation of the considerable body of knowledge and experience which already exists.

This knowledge and experience is not unique to any one group involved in the specification, application, and operation of mechanical seals. A cooperative effort between the interested parties was, therefore, essential to make the best use of existing information. For this reason, an industry working party was established by the Institution of Mechanical Engineers with representation from seal and pump manufacturers, users, contractors, and research organisations, with the remit of proposing ways of achieving more consistent seal performance.

This working party was formed in July 1984. It quickly became evident that, although no single cause could be attributed as the principal factor in seal failure, a set of guidelines indicating the main prerequisites for reliable seal performance could be an important first step leading to early benefits. This book is the outcome. The various chapters have been written by specialists in the different fields and, while they generally reflect the background of the individuals concerned, they have been subject to intensive scrutiny and revision by the other interests on the working party. In this way, they represent a wider point of view as the result of genuine and understanding cooperation. It must be emphasized that these are guidelines, not standards as such, since standards are not part of the Institution's work. Improved standards are certainly required. For many years, a mechanical seal was considered as an alternative to gland packing. No account was taken of the requirements of the seal in designing the stuffing box. This is now, fortunately, changing. There are other recommendations in this book which, it is hoped, could be included in future standards, particularly relating to design, manufacture, testing, and provision of information to the various parties involved in mechanical seal application.

There is evidence, from studies of failure statistics, that seal lives in excess of the average 8–13 months are possible, up to, in some cases, 3–5 years. However, even in industries where quality assurance, in every sense of the term, is practised, for example, nuclear power, unexplained 'infant mortality' and random failures do occur. In the present state of knowledge, a general rule which quantifies expected levels of improvement from specific courses of action does not appear practical. A study of the principal causes of seal failure, however, does indicate how this guide will be of value in assisting improvement in current average levels of performance. Furthermore, recent experience reported in the USA, where attention has been paid to the topics discussed in this guide, provides additional grounds for confidence.

Surveys of pump populations carried out over a period of years in the USA suggest that the basic causes of seal failure and their relative importance are as follows.

Operating problems (flush flow interruption, etc.)	40 percent
Mechanical difficulties (assembly errors, incorrect clearances, inadequate alignment, etc.)	24 percent
Faulty fluid circuit design (insufficient suction, deficient flush circuit, pump characteristics unsuitable for service)	19 percent
Seal component selection inadequate (materials or basic configuration in error)	9 percent
Miscellaneous causes	8 percent

All the above topics are addressed in this guide. The major causes of failure are for the most part beyond the scope of the seal designers' responsibility, springing from possible lack of knowledge and training; certainly from a lack of real appreciation of those factors which can clearly cause failure.

The other major component in a pump which can cause an outage is the bearings; in most cases these are rolling contact bearings. The mean time between failures of these components on average is probably somewhat in excess of 3 years. An L_{10} life for such bearings is commonly specified as 25 000 hours continuous operation. The concept of an L_{10} life is not really appropriate to the case of seals, where failures are not yet predominantly the result of wear. However, a mean time between failures of three years might reasonably be considered as a target for seal life, reducing the gap between seals and bearings as potential sources of pump outage. This, on current evidence, may well be feasible as a first step towards reducing pump outage with considerable economic benefit.

It is almost certain that this is the first time that mechanical seals have been studied in this way and the result is a unique and practical contribution to their application. It is considered that, through a real effort to appreciate the other parties' point of view, this book must be of interest and value to seal designers, pump manufacturers and users alike.

Mechanical seal technology continues to develop rapidly. It may, therefore, prove necessary to produce revisions in future years and any reader comments, particularly from users, would be much appreciated by the Working Party to assist in this task. The Working Party can be contacted through the Institution of Mechanical Engineers, 1 Birdcage Walk, London SW1H 9JJ.

The members of the working party, together with the Companies which they represent, are listed on the title page. Their participation in this book is gratefully acknowledged, as indeed is the support provided by their parent companies.

A. G. H. Coombs

EDITOR'S COMMENTS

Units

A complete notation is given in Appendix 2.

The equations derived in the text are expressed coherently and can thus be used with any consistent system of units.

Empirical equations and tables taken from published sources are given in the units in which they were derived in the original. Different symbols are used in these equations and the relevant units indicated both in the text and in Appendix 2.

Figure 3.2 and Table 6.1 use symbols already well established in practice. They differ from those generally used in the text and are not listed in Appendix 2.

Proprietary and Trade Names

The following trade names appear in the text. They are distinguished by an initial capital letter.

Carboloy 833	General Electric Co., USA
Carpenter 20 Cb3	Carpenter Technology Corporation
Colmonoy	Wall Colmonoy
Fluorel	3M Company
Freon	E. I. Du Pont de Nemours
Grafoil	Union Carbide
Graftite	Crane Packing Limited
Hastelloy	Cabot Corporation
Inconel	International Nickel Company
Kalrez	E. I. Du Pont de Nemours
Monel	International Nickel Company
Refel	Turner and Newall
Stellite	Cabot Corporation
Viton	E. I. Du Pont de Nemours

PART I

Mechanical seal design

Chapter 1

BASIC CONCEPTS OF SEAL FUNCTION AND DESIGN
N. M. Wallace

1.1 PRINCIPLES OF OPERATION

1.1.1 General description

Mechanical seals are the most versatile type of seal for rotating shafts. Their main use is on liquid/gas sealing, e.g., centrifugal pumps. They are also used on gas/gas applications, e.g., compressor and agitator shaft seals, but in these cases they are usually deployed as double seals with liquid injection to provide lubrication of the seal faces. Liquid/liquid sealing duties occur in such applications as double seals where process fluid is on one side and barrier fluid on the other.

Dry-running gas seals have recently been developed and are used for such applications as compressor and turbine shafts. They are outside the scope of this publication, as are dry-running back-up seals.

Mechanical seals have largely replaced soft-packed glands because of low leakage and the absence of a requirement for routine maintenance.

Mechanical seals in principle consist of two plane faces arranged perpendicular to the axis of a rotating shaft. This gives rise to the alternative name of 'radial face seal'. One face is fixed to the pump casing or vessel and is stationary, the other is fixed to the shaft and rotates with it.

In order to keep leakage to an acceptable level, it is necessary that the two faces run with very small separation, typically less than 0.001 mm.

In order to keep frictional heat generation and wear within acceptable limits, a lubricating film of liquid must be maintained between the seal faces, while not exceeding the film thicknesses mentioned above. In most cases, where single seals are employed, this lubricant will be the sealed fluid, which may have poor lubricating properties.

In order to accommodate wear and pump build tolerances, it is necessary that one of the faces can move axially and that one of the faces has the freedom to accommodate swash (angular misalignment in the rotating component). Usually the same face has the freedom to accommodate both these movements.

The floating face may be either stationary or rotating, but for high speed applications (shaft surface speed greater than about 15 m/s) stationary mounting avoids problems arising from centrifugal forces.

Normally the seal is mounted 'internally', with the sealed fluid on the outer periphery of the seal. With highly corrosive liquids, however the seal can be mounted 'externally' to minimize exposure of the seal parts to the corrosive conditions.

Mechanical seals thus have a superficial resemblance to thrust bearings. However, the very close running clearance which must be maintained, poor lubricating properties of the 'lubricant', and very low lubricant flow combine to make the design of mechanical seals much more difficult than that of thrust bearings.

The comparison of the main operating parameters in mechanical seals and thrust bearings, given in Table 1.1, highlights the problems inherent in the successful and reliable operation of mechanical seals.

Table 1.1. Comparison of the main operating parameters of mechanical seals and oil lubricated hydrodynamic thrust bearings

	Mechanical seal	Thrust collar
Specific face loading (MPa)	0.5–15	≤0.5
'Lubricant' viscosity (Ns/m²)	0.0005–0.1	0.01–0.015
'Lubricant' flow	ml/hr	l/min
'Lubricant' film thickness (μm)	0.5–2.0	10–15
Surface flatness (μm)	<1	<10
Surface finish, Ra (μm)	0.1–0.2	0.2–0.4

The problems in designing mechanical seals explain why the reliability is generally worse than for thrust collars.

It is the intention of this chapter to provide an introduction to the factors which must be considered in the design of seals. This will permit the user to engage in constructive dialogue with the seal and pump vendors and to appreciate and exploit those things which will contribute to increased reliability. It is not the intention of this chapter to provide all the information necessary to design seals.

Mechanical seals are precision devices, more expensive in initial and replacement costs than soft-packed glands and, if reliable performance is to be achieved, require careful attention to detail in design of the seal arrangement, together with high standards of fitting and proper operation.

Mechanical seals are produced in a variety of designs, sizes, and materials to meet the varied requirements of users. Catalogue product ranges have been evolved by individual seal manufacturers to meet the majority of these needs, and the technology exists to meet other difficult or unusual applications.

A typical mechanical seal installation in a centrifugal pump is shown in Fig. 1.1. This shows the basic terminology of the main seal components; a more complete explanation of the terms used in mechanical seal technology is given in Appendix 1.

The seal shown in Fig. 1.1 is a simple light-duty design. The sealing interface where the seal rotating and stationary faces make rotational contact is a plane annulus. The sealing faces themselves are lapped to a very high degree of flatness, measured in wavelengths of light. The body of the rotating component is free to float axially. This enables hydraulic pressure to keep the main sealing faces together in operation; in the absence of hydraulic pressure the spring serves the same purpose.

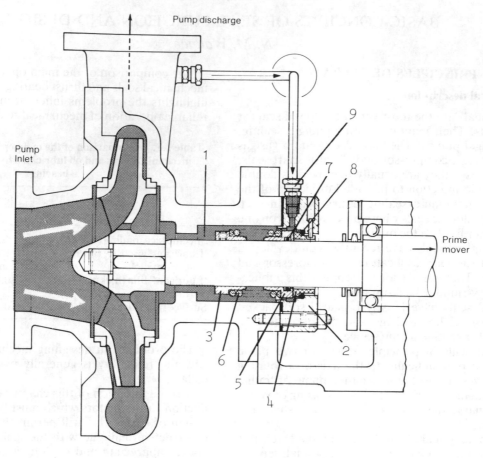

Fig. 1.1.   Typical mechanical seal installation
(1)   **Seal chamber**
(2)   **Seal plate**
(3)   **Spring sleeve**
(4)   **Seat**
(5)   **Seal ring**
(6)   **Spring**
(7)   **Dynamic secondary seal**
(8)   **Static secondary seal**
(9)   **Flush connection**

The closing force on the seal is designed to provide an acceptable compromise between leakage and wear. It must also be sufficient to prevent the faces from separating in the event of mechanical shock or a pressure fluctuation. A secondary seal prevents leakage between the floating assembly and the shaft; often a rubber 'O' ring is used for this purpose.

Friction arising from sliding at the sealing interface causes considerable heat generation, typically between 100 and 1000 watts for seals of modest size and speed. This heat is dispersed by conduction through the body of the seal and thence into the surrounding liquid. Leaked fluid removes only an insignificant amount of heat.

It is important that the interface temperature is held safely below the boiling point of the sealed fluid. If the temperature of the fluid surrounding the seal is too high for effective cooling it is usual to arrange for injection of cooled fluid into the seal chamber or to provide other supplementary cooling facilities.

Materials for seal faces are chosen for good sliding performance (bearing compatibility), resistance to attack by the sealed fluid (chemical compatibility), and compatibility with temperature and any other operational factors (e.g., abrasion, radiation, etc.).

Usually one face is a carbon–graphite material and the mating face a metal alloy, ceramic, or carbide. Carbon–graphite combines exceptional chemical compatibility and bearing properties; when used as sealing rings, carbon–graphites commonly contain a metal or resin impregnation which must also be compatible.

Lubrication of the sliding interface is designed to be self-maintaining. Depending on duty and seal design, it is provided by a fluid film separating the faces, or by a 'combined' fluid film/asperity contact process; either way it must be effective with the sealed fluid as the 'lubricant'. Satisfactory seal performance depends on effective lubrication of the sliding interface.

1.1.2 Design variants

Many variants of the simple design shown in Fig. 1.1 have evolved to cope with the wide variations in shaft size, speed of rotation, sealed pressure, temperature, and fluid properties that occur in practice, not only in centrifugal pumps, the main field of application, but also to cover other applications such as compressors, agitator shafts, etc. These include the following.

Seal configurations
 For example, internally and externally mounted; internally and externally pressurized.

Different types of secondary sealing elements
 PTFE wedges, 'U' cups, trapezoidal packings, etc.

Different spring arrangements
 Single or multi springs in rotating or stationary configurations.

Bellows seals in metals or rubber
 Bellows seals eliminate the secondary sliding seal and metal bellows also eliminate the spring.

Alternative materials of construction
 Faces can be in monobloc, hard faced, or inserted constructions in a wide variety of materials. A wide variety of materials is used for other components.

Different load/balance configurations
 Mechanical seals are available in unbalanced and balanced form. The degree of balance in both types can vary, as will be illustrated later.

Whatever the particular construction, the basic objective in radial face mechanical seal design is to achieve a stable fluid film condition at the seal face such that leakage is limited to acceptable levels while achieving the desired life.

The aim of this chapter is to give an insight into the important parameters in design and highlight the ways in which the performance of mechanical seals can be improved.

1.2 FACTORS AFFECTING DESIGN

The equations developed in this chapter are generally for seals pressurized on their outer periphery. The equations for seals pressurized at their inner periphery follow the same principles.

1.2.1 Closing Force: F_t

The total closing force on the floating member of a seal, F_t is the sum of

 The spring load, F_s
 The hydraulic load, F_h
$$F_t = F_s + F_h \tag{1.1}$$

The specific closing force p_n is based on the area of the sealing interface, A_f

$$p_n = F_t/A_f$$
where
$$A_f = \tfrac{1}{4}\pi(D_o^2 - D_i^2)$$

D_o and D_i are the outside and inside diameters of the sealing interface.

The closing force is reacted in the sliding interface by a supporting force which is provided by some combination of

 Hydrostatic fluid pressure
 Hydrodynamic fluid pressure
 Solid or asperity contact

These forces are discussed further in section 1.2.2 under Lubrication.

Spring load: F_s

All mechanical seals have an arrangement (spring or metallic bellows) to apply an initial closing force to the faces and to hold them together in the absence of fluid pressure (Fig. 1.2). The magnitude of the spring load must be sufficient to overcome any axial friction from the dynamic secondary seal, the dynamic effects of any face misalignments and, for general purpose seals, must allow for the possibility of a vacuum in the seal chamber which, in effect, subjects the seal to a reverse pressure. (*Note.* It is also possible to over-pressure a single mechanical seal in the wrong (reverse) direction if it is fitted with an atmospheric quench and the quench pressure is too high. Similarly with tandem and double mechanical seals it is possible to generate reverse pressures which can displace secondary seals such as 'O' rings. This is at least inconvenient, since the seal must be dismantled to replace them, and at worst potentially dangerous, since a major leakage could occur. Major leakages are possible when the primary seal pressure is re-established in the case of tandem seals or the barrier fluid pressure is lost in the case of double seals. Designs which are tolerant to transient reverse pressure effects are available.)

Referring to Fig. 1.2

 D_b = balance or sliding diameter
 D_h = outer diameter of stationary secondary seal

The face pressure due to the spring, p_s, also depends on the area of the sliding interface

$$p_s = F_s/A_f \tag{1.2}$$

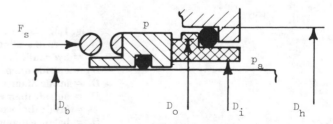

Fig. 1.2. Arrangement of spring in mechanical seal

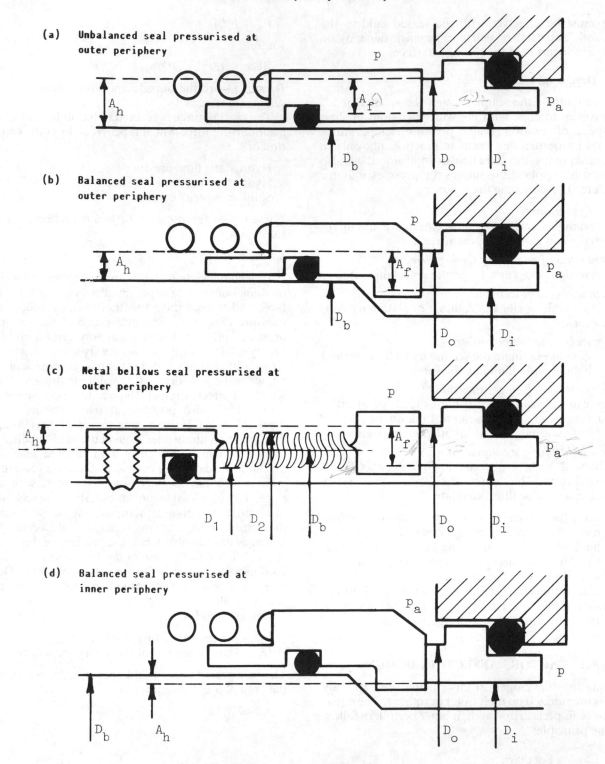

Fig. 1.3.   Balance in mechanical seals
A_f = area of sealing interface
A_h = hydraulic loading area
D_2 = outside diameter of bellows
D_1 = inside diameter of bellows
D_o = outside diameter of sealing interface
D_i = inside diameter of sealing interface
D_b = balance diameter

Typical values of spring load per unit area of sealing interface, (p_s), are 0.1 to 0.3 MPa, though in some designs values as low as 0.02 MPa or as high as 0.5 MPa may be used.

to resist vacuum loads

$$F_s > \tfrac{1}{4}\pi(D_n^2 - D_b^2)(p_a - p)$$

or if the stator is clamped

$$F_s > \tfrac{1}{4}\pi(D_o^2 - D_b^2)(p_a - p)$$

where

p = pressure on outer periphery of seal
p_a = pressure on inner periphery of seal

Balance ratio: B

A fundamental characteristic of a mechanical seal is it's 'balance ratio'.

Seals in which the balance ratio is equal to (or greater than 1) ar referred to as unbalanced seals; those in which it is less than 1 as balanced seals. Unbalanced seals are used for lower pressures, normally 1 MPa g (10 bar g) max. in simple applications; balanced seals are used for higher pressures, typically up to 7 MPa g (70 bar g) with general purpose balanced seals, and higher with duty designs.

All mechanical seals contain an area within their design which is responsible for generating the hydraulic closing forces in the seal. This area is annular, enclosed by the outside diameter of the sealing face (D_o) and the balance diameter (D_b). It is the area of the seal face compared to this hydraulic area which determines whether a seal is balanced or unbalanced and this, in turn, determines the prevailing bearing pressure at the faces and the type of fluid film formed and, hence, the seal's ultimate pressure capability.

The balance ratio, B, can be derived from the following formulae

for an externally pressurized seal

$$B = \frac{\text{hydraulic loading area}}{\text{sealing interface area}} = \frac{\tfrac{1}{4}\pi(D_o^2 - D_b^2)}{\tfrac{1}{4}\pi(D_o^2 - D_i^2)}$$

$$= \frac{(D_o^2 - D_b^2)}{(D_o^2 - D_i^2)} \quad (1.3)$$

for an internally pressurized seal the hydraulic load area is $\tfrac{1}{4}\pi(D_b^2 - D_i^2)$

$$B = (D_b^2 - D_i^2)/(D_o^2 - D_i^2) \quad (1.4)$$

Figure 1.3 shows that balanced seals have face areas which are larger than the hydraulic loading areas by disposing the face area above and below the balance line as indicated. This requires a step in the shaft or sleeve; whereas, in the case of unbalanced seals, the face area is smaller than the hydraulic area; unbalanced seals may be applied directly to a parallel shaft.

The balance ratios of commercially available balanced seals vary typically from 0.65 to 0.85. The choice of balance results from the compromise effected by the particular seal manufacturer. The choice may be influ-

enced by the seal type and its characteristics of design and a compromise between the opposing requirements of high sealing integrity and controlled face loading.

Seals with a high balance ratio, e.g., $B = 0.85$, are more stable and less likely to blow open under the action of hydrostatic pressure between the faces, but at the expense of higher face loadings. Equally, seals with a lower balance ratio, e.g., $B = 0.65$, exhibit lower face loadings and lower amounts of heat generated, but can be hydraulically unstable at higher pressures if not designed carefully.

Unbalanced seals typically have balance ratios of around 1.2; the degree of unbalance depends on the clearance between the shaft and the seal face inner circumference.

Bellows seals achieve balance without a step in the shaft since the 'O' ring at the sliding diameter is eliminated and the balance or sliding diameter (D_b^2) is 'built-in' to the bellows. The sliding or balance diameter lies between the inside and outside diameters of the bellows $(D_1$ and $D_2)$ and, for metal bellows, is normally calculated as the root mean square of those values

$$D_b = \sqrt{(0.5(D_2^2 + D_1^2))} \quad (1.5)$$

In seals with metal bellows, the spring function is normally performed by the bellows itself, whilst rubber and plastics bellows usually incorporate a metallic spring.

Balance ratios are calculated as previously described, using the balance diameter in conjunction with the seal face diameters.

Hydraulic load: F_h

Seals are normally subjected to different pressures on the inside and outside. In effect, the seal behaves like two seals: one internally and the other externally pressurized (Fig. 1.4).

The hydraulic closing force may be obtained by superposition

$$\begin{aligned}
F_h &= \tfrac{1}{4}\pi(D_o^2 - D_b^2) \times p + \tfrac{1}{4}\pi(D_b^2 - D_i^2) \times p_a \\
&= \tfrac{1}{4}\pi(D_o^2 - D_i^2) \times B \times p \\
&\quad + \tfrac{1}{4}\pi(D_o^2 - D_i^2)(1 - B) \times p_a \\
&= A_f \times B \times p + A_f(1 - B) \times p_a \\
&= A_f(B \times p + (1 - B) \times p_a) \quad (1.6a) \\
&= A_f(B \times \Delta p + p_a) \quad (1.6b)
\end{aligned}$$

where Δp is the pressure differential across the seal. (Note that for an unbalanced seal, $B > 1$, the pressure p_a acts to reduce the closing force.)

Total closing force: F_t

The total closing force in a mechanical seal is thus very predictable and depends on the initial spring load and the hydraulic closing force.

Total closing force (F_t)
= Spring force (F_s) + Hydraulic closing force (F_h)
$$F_t = F_s + A_f(\Delta p \times B + p_a) \quad (1.7)$$

It is the way in which this load is supported that is not so predictable, as will be described in later sections.

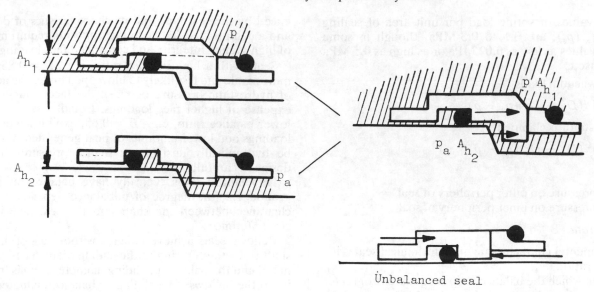

Fig. 1.4. The elements of hydraulic load in a seal

For very high pressure ($p > 30$ bar) seals, sealing to atmosphere ($p_a = 1$ bar a), p_a may be neglected, and equation (1.7a) simplifies to

$$F_t = F_s + p \times B \times A_f \qquad (1.7b)$$

Examples are given wherever possible in this chapter and the 70 mm size seal used here appears regularly throughout.*

Example 1.

A balanced seal (sealing to atmosphere), 70 mm size, sealed pressure: 1.1 MPa (1 MPa g)

$$p_a = 0.1 \text{ MPa} \qquad\qquad D_i = 80.17 \text{ mm}$$
$$p = 1.1 \text{ MPa} \qquad\qquad D_b = 82.45 \text{ mm}$$
$$\Delta p = 1.0 \text{ MPa} \qquad\qquad F_s = 285 \text{ N}$$
$$D_o = 87.17 \text{ mm}$$

$$B = (D_o^2 - D_b^2)/(D_o^2 - D_i^2)$$

$$= (87.17^2 - 82.45^2)$$

$$\div (87.17^2 - 80.17^2)$$

$$= 0.683$$

$$A_f = \tfrac{1}{4}\pi(87.17^2 - 80.17^2)$$

$$= 920 \text{ mm}^2 \ (1.43 \text{ in}^2)$$

$$F_t = 285 + 920(1 \times 0.683 + 0.1)$$

$$= 1005.4 \text{ N } (226.2 \text{ /lbf})$$

Example 2.

An unbalanced seal, 70 mm dia, with the same face area, spring load, and duty conditions. Typical value of $B = 1.2$.

---

*In the examples in this chapter, seal size is based on the maximum shaft or sleeve which will pass freely through the seal. Different conventions are used by different manufacturers and in many cases, the balance diameter itself is the basis for sizing the seal.

Then

$$F_t = 285 + 920(1 \times 1.2 + 0.1)$$

$$= 1481 \text{ N } (333.2 \text{ lbf})$$

The above examples show the fundamental difference in face load between balanced and unbalanced seals. This is not, however, entirely responsible for their different pressure capability, as will be explained in section 1.2.2 under Film formation.

1.2.2 Lubrication

Some form of lubrication between the seal faces is necessary to control frictional heat generation and wear of the faces. Three modes are possible:

Hydrostatic film

Hydrodynamic film

Boundary lubrication

In hydrostatic lubrication the faces are separated by the pressure of the sealed fluid. The pressure in the separating film can also be generated by hydrodynamic action resulting from the relative movement of the faces, but the conditions with normal seal faces are not favorable for hydrodynamic action and any film that is formed tends to be very thin. Seal faces can be designed specifically to generate thicker hydrodynamic films; this results in safer operating conditions but at the cost of greater leakage. These two modes are covered by the general term, fluid film lubrication, where the faces are separated but the mechanism of pressure generation in the film is not specified.

It is regularly observed that general purpose seals exhibit distinct and repeatable wear rates when tested on clean fluids. This is indicative of asperity contact across the fluid film and an element of mechanical load support. In contrast, some seals on very high duty conditions exhibit no wear or evidence of face contact due to the full film support of their fluid films which are generated either hydrodynamically, hydrostatically, or a combination of the two. Indeed asperity contact in some seals could initiate a change of regime and possibly failure following a loss of lubricity through overheating of the film.

Pure hydrostatic operation can result in low wear rates but relatively high leakage rates; it tends to be associated with the highest levels of reliability.

Hydrodynamic operation produces performance intermediate to the other two categories. Hydrostatic and hydrodynamic lubrication are governed by well-defined equations of fluid dynamics, given in standard text books on lubrication theory.

The total closing (F_t) force in a mechanical seal was shown (equation 1.7a) to be

$$F_t = F_s + A_f(\Delta p \times B + p_a)$$

The whole of this closing force is supported by some combination of hydrostatic film support (F_{hs}), hydrodynamic film support (F_{hd}), and mechanical support (F_m) through asperity contact.

$$F_t = F_{hs} + F_{hd} + F_m \qquad (1.8)$$

In the rest of this section we will look at the various mechanisms and the levels of load support that can be expected. The effect of the variables available in seal design will also be assessed in an attempt to show what can be done to optimize performance whilst retaining a high degree of stability.

Hydrostatic lubrication

Hydrostatic lubrication depends on the pressure of the sealed fluid to create a pressure field to support the load on the interface, and it is sensitive to face geometry, particularly 'coning' (axisymmetric radial variation in film thickness).

The hydrostatic load support is derived from the pressure distribution across the seal face, where clearly,

the pressure must fall in some way from the sealed pressure on the fluid side to the pressure on the other side (atmospheric pressure in the case of a single seal), Fig. 1.5.

The pressure distribution is dependent on the shape of the faces and, in particular, their conicity. Whilst it is clear that in many seals the face shape will not be purely conical, but may vary around the circumference through waviness of the faces, it is interesting to look at the effect of the conical shape of the face on 'load sharing'.

The pressure distribution factor, β, is a very convenient parameter in the calculation and illustration of hydrostatic load support. It is defined in equation (1.9)

$$\begin{aligned} F_{hs} &= \int_{R_i}^{R_0} 2\pi r \, \delta r \, p(r) \\ &= p_a \times A_f + \beta(p - p_a)A_f \\ &= A_f(\beta \times \Delta p + p_a) \qquad (1.9) \end{aligned}$$

In effect, β is a factor which integrates the distribution of the pressure differential across the face so that an average film pressure of $\Delta p \times \beta + p_a$ may be determined for use in calculating the opening force at the face.

In some very high pressure seals, hydrostatic load support is vital, particularly where two hard faces are employed. A transition to boundary or mixed lubrication could lead to severe surface distress at the seal faces (e.g., thermal-stress cracking) and probable breakdown of the seal itself.

Residual load support

The total closing force, F_t, on a seal was defined (equation 1.7a) as

$$F_t = F_s + A_f(\Delta p \times B + p_a)$$

and

$$F_t = F_{hs} + F_{hd} + F_m$$

Hence the residual load support, $F_{hd} + F_m$, may be determined

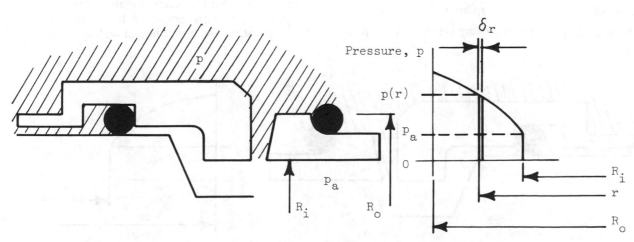

Fig. 1.5. Seal face pressure distribution

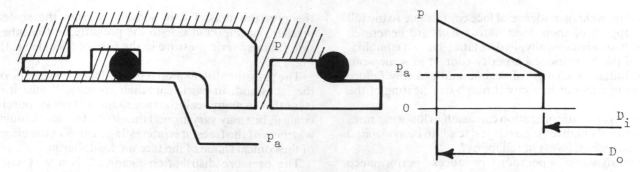

Fig. 1.6.   Linear pressure drop across parallel seal faces

$$F_{hd} + F_m = F_t - F_{hs}$$
$$= F_s + A_f(\Delta p \times B + p_a)$$
$$\quad - A_f(\Delta p \times \beta + p_a)$$
$$= A_f \times p_s + A_f(\Delta p \times B + p_a)$$
$$\quad - A_f(\Delta p \times \beta + p_a)$$
$$= A_f(p_s + \Delta p \times B + p_a - \Delta p \times \beta - p_a)$$
$$= A_f(p_s + (B - \beta)\Delta p) \qquad (1.10)$$

The residual closing force is often referred to as the net closing force, F_{net}.

The net closing pressure, p_f, may be defined as F_{net}/A_f. In many cases, the hydrodynamic load support, F_{hd}, is negligible and all the residual load is taken by F_m, in asperity contact.

We will now look at the effect of the important element of hydrostatic load support, F_{hs}, in parallel, converging and diverging fluid films.

Parallel films

The most common assumption made is that the pressure falls linearly across the face of a mechanical seal through a parallel fluid film (Fig. 1.6). In this case $\beta = 0.5$.

From equation (1.9)

$$F_{hs} = A_f(\Delta p \times \beta + p_a) = A_f(\Delta p \times 0.5 + p_a)$$

and the average pressure in the film is $0.5 \times \Delta p + p_a$

Example 3.

From the previous example with a balanced seal ($B = 0.683$) at 1 MPa g and a spring load of 285 N the total closing force, F_t, was calculated at 1005.4 N. Face area was 920 mm^2. Assuming F_{hd} is negligible, then

$$F_m = 920 \times \{285/920 + 1(0.683 - 0.5)\}$$
$$= 453.36 \text{ N}$$

i.e.

$$F_m/F_t = 453.36/1005.4 = 0.451$$

In this example, the mechanical load support thus accounts for 45 per cent of the total load.

Converging films

With the faces converging in the direction of sealing, the pressure distribution becomes non-linear (Fig. 1.7) and β becomes greater than 0.5, hence the average pressure becomes greater than $\Delta p/2 + p_a$. In the limit with the faces in contact at the inner diameter there is no leakage and $\beta = 1$. This can be an unstable condition as the separating force may exceed the closing force and the faces can begin to separate. The act of separating the faces creates leakage and this causes the pressure distribution to reduce; infinite separation would give $\beta = 0.5$. An equilibrium condition exists when the prevailing pressure distribution exactly balances the closing force. As β increases with increasing face convexity, greater face separation and hence higher leakage result.

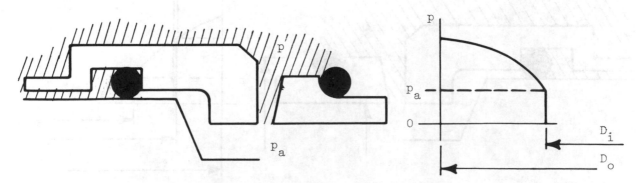

Fig. 1.7.   Pressure drop across a converging film

It is possible to define the minimum differential pressure Δp_o at which face separation can occur by equating the opening and closing forces F_{hs} and F_t with a pressure distribution factor β of 1. That is

$$F_{hs} = F_t \quad \text{and} \quad (F_{hd} + F_m) = 0$$

By substituting in equation 1.10,

$$p_s + \Delta p(B - \beta) = 0$$

With $\beta = 1$, the minimum differential opening pressure, Δp_o, is given by

$$\Delta p_o = p_s/(1 - B) \tag{1.11}$$

Similarly it is possible to define the critical value of β, (β_c) at a particular pressure. This is the maximum value of the pressure distribution factor before separation can occur with a pressure differential of Δp. Substituting again in equation 1.10,

$$p_s + \Delta p(B - \beta_c) = 0$$

and

$$\beta_c = (p_s/\Delta p) + B \tag{1.12}$$

Example 4.

Taking the 70 mm balanced seal given in Example 1 ($F_s = 285\,\text{N}, A_f = 920\,\text{mm}^2$) we get the following results

'blow-open' pressure differential

$$\Delta p_o = p_s/(1 - B)$$
$$p_s = 285/920 \times 10^{-6}\,\text{Pa} = 0.309\,\text{MPa}$$
$$\Delta p_o = 0.977\,\text{MPa}\ (142\,\text{lbf/in}^2)$$

Clearly, there are many balanced mechanical seals which are operating in a regime in which they could potentially blow open and leak excessively. The reason that they do not is a matter of design, where excessive conicity is avoided and hence the value of β is limited. Most seals 'run in' to a stable equilibrium condition with minimal leakage.

In fact many seals bed in to give a parallel face condition and operate in a mixed lubrication regime.

Example 5.

The critical value of pressure distribution β_c.
Taking the same seal with a sealed presure of 1.5 MPa g (217 lbf/in^2 g)

$$\beta_c = (p_s/\Delta p) + B$$
$$= 0.309/1.5 + 0.683$$
$$= 0.889$$

At higher pressures, the permissible value of β falls. The equivalent value at 15 MPa g (2175 lbf/in^2 g) is only 0.704.

Section 1.2.5 on Seal Face Deflections illustrates how the flatness of seal faces can be influenced and serves to show that seal design needs to be good to take advantage of low balance ratios at high pressures, whilst retaining stability.

An approach to seal stability

In the context of seal face stability, it is interesting to consider one practical approach to its achievement. The practice relates to the sealing of light hydrocarbons, where the dangers of leakage are high and it is possible to get very high values of β at the faces due to hydrostatic effects and additionally due to vaporizing film effects.

The objective is to ensure that, at the operating pressure, there is always a positive net closing pressure (p_f) at the face of between 0.05 and 0.35 MPa (7.3 and 50.8 lbf/in^2), assuming full pressure across the whole of the face ($\beta = 1$). That is

$$0.05 < p_f < 0.35\,\text{MPa}$$

In this way it is theoretically not possible for the seal to blow open hydrostatically. The requirement is met by combining smaller face areas and/or less balanced seals (often $B > 0.9$), but much care must be taken so that the general running ability of the seal is not impaired.

If, however, the value of β is less than 1, say $\beta = 0.5$, then the actual net pressure at the face will be much higher than it would have been in a more conventional design.

By definition

$$p_f = F_{net}/A_f = (F_{hd} + F_m)/A_f$$

from equation 1.10

$$F_{hd} + F_m = A_f\{p_s + \Delta p(B - \beta)\}$$

thus

$$p_f = p_s + \Delta p(B - \beta)$$

with $\beta = 1$

$$p_f = p_s + \Delta p(B - 1)$$

whence

$$B = (p_f - p_s)/\Delta p + 1$$

Example 6.

Take the 70 mm seal at $p = 4\,\text{MPa g}$ (580 lbf/in^2 g) and set $p_f = 0.05\,\text{MPa}$ (7.3 lbf/in^2),

$$B = (0.05 - 0.309)/4 + 1$$
$$B = 0.935$$

However if $\beta = 0.5$, $P_f = 0.309 + 4(0.935 - 0.5) = 2.05\,\text{MPa}$ (297 lbf/in^2) and heavy face loading results.

Diverging films (Concave faces)

In order to ensure that balanced seals do not leak when pressure is applied for the first time, they are often arranged so that any small deflections due to pressure or thermal effects produce concave faces (Fig. 1.8). In this case, the rate of pressure drop is greater than with a parallel film (see Fig. 1.6) and $\beta < 0.5$.

In fact, this normally means that there is initially no fluid film. Contact is at the outer periphery and is heavy. Given the right materials of construction, the faces will then bed-in and eventually result in a parallel film with a linear pressure distribution.

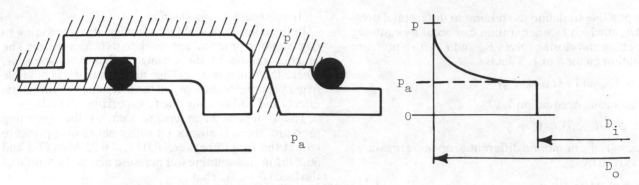

Fig. 1.8.   Pressure drop across a diverging film

Initial running friction will be high and largely dependent on the dry values of the face materials. These values will fall during the running in process and stabilize at a value which is largely determined by the eventual operating regime.

Given this bedding in mechanism, it is clear why carbon is chosen as a seal face material in so many seals. It has the ability to act effectively as a dry bearing and wears so that the two faces can conform. With seals with two hard faces (e.g., tungsten carbide), these forgiving characteristics do not exist and concave faces can produce heavy thermal cracking around the outer periphery at significant operating pressures. At best, such seals will run with high friction levels and at worst could destroy themselves in the attempt to run in. Hard faced seals require more care in design. The concave deflections in carbon faced seals must also be limited, so as not to produce excessive leakage on reduction of pressure (see section 1.2.5).

Hydrodynamic lubrication

Hydrodynamic lubrication depends on relative movement between the seal faces. This can involve both circumferential movement with faces exhibiting waviness, or squeeze-film effects from oscillating axial movements. Both these effects can produce a net positive pressure in the interface film. These mechanisms are influenced by the fluid viscosity, the face width, flatness, and conicity, in addition to the speed of rotation. While, in principle the mathematical analyses are well developed, it is difficult to apply them in practice, as much of the information required is not well defined in the case of seal faces.

It is generally accepted that hydrodynamic effects play a part in the operation of most mechanical seals. The contribution to load support may vary from small to significant. Seals are available with slotted faces designed to give increased hydrodynamic effects. These can give enhanced reliability and life, but at the expense of increased leakage. Care must be taken where abrasives are present since the increased film thicknesses with this type of seal can allow abrasive particles to penetrate between the faces and create excessive wear.

The subject of hydrodynamic lubrication in mechanical seals is too complex to be treated in greater detail in

this guide; additional information is given in the references listed in the Bibliography (Appendix 3, section 3.3, Nau (1985), section 5.4).

Boundary lubrication

Boundary lubrication depends on the formation of adsorbed films on the contacting surfaces and is thus influenced by the chemical nature of both the surfaces and the fluid in contact with them. In general, most of the fluids sealed by mechanical seals (aqueous solutions, light hydrocarbons) have poor boundary lubricating properties and this imposes limitations on the materials that can be used for the faces.

Self-lubricating materials, such as carbon graphite and filled PTFE, are good in this respect, but they are soft and tend to wear in the presence of abrasive solids. Hard faces are effective in resisting wear, but have to be selected carefully if they are to perform satisfactorily under conditions of solid contact, particularly in the case of unbalanced seals that at high pressure operate substantially in the boundary lubrication mode.

Typical values of the coefficient of friction of mechanical seal face material combinations running under dry conditions are given in Table 1.2.

These provide some guidance on material selection and particularly on the amount of heat generation, though the ultimate choice must be based on the performance of the material in contact with the sealed

Table 1.2.   Typical coefficients of friction for mechanical seal face pairs

Material combination	Friction coefficient (dry)
Metal/PTFE	0.1 variable with load and speed
Tungsten carbide/carbon	0.1–0.15
Silicon carbide/carbon	0.1–0.15
Stellite/carbon	0.2–0.25
Tungsten carbide/tungsten carbide	0.2
Silicon carbide/silicon carbide	0.25
Tungsten carbide/silicon carbide	0.2

Notes. The above values represent an average of many quoted values. Those for silicon carbide are for reaction-bonded grades.

liquid, mainly a matter of experience. Material selection is discussed in more detail in Chapter 2.

Film formation

Although, as we have seen in the previous section, it is difficult to predict the precise lubrication conditions existing between seal faces, some fluid film formation is desirable to reduce wear of the faces to an acceptable level. It is thus worth looking at two effects that tend to inhibit fluid film formation.

Specific face pressure effects (parallel faces)

As previously shown, mechanical seals are closed by a predictable force F_t. This force is reacted over the area of the face A_f. It is possible to ascribe a value, p_n (see section 1.2.1), to the specific closing pressure at the face. If the faces are in reasonable conformity, this value, p_n, is the load bearing pressure at the face itself. If the value of p_n is sufficiently high it will tend to squeeze the film out from between the faces.

$$p_n = F_t/A_f = \{F_s + A_f(\Delta p \times B + p_a)\}/A_f$$
$$= p_s + \Delta p \times B + p_a$$

The forcing pressure to produce the film is the pressure p and, clearly, if $p > p_n$, then the film will form readily, but if $p < p_n$, then film formation will be inhibited.

Define $\delta p = p - p_n$

$$\delta p = p - p_s - \Delta p \times B - p_a$$
$$= \Delta p(1 - B) - p_s \qquad (1.13)$$

if $\delta p > 0$ the film will form readily
if $\delta p < 0$ film formation will be inhibited

If $\Delta p > p_s/(1 - B)$, then $\delta p > 0$ for balanced seals ($B < 1$) and $\delta p < 0$ for unbalanced seals ($B > 1$).

An example of the variation of δp with Δp is presented graphically in Fig. 1.9.

The first observation that can be made from the graphs in Fig. 1.9 is that both balanced and unbalanced seals have an equal ability to form a film at zero pressure, since both have the same value of δp (viz p_s). As the pressure rises, film formation in the balanced seal becomes more favorable, whereas, with an unbalanced seal, it becomes progressively worse.

Interestingly the value of δp is zero when $\Delta p = p_s/(1 - B)$ i.e., $\Delta p = \Delta p_o$ (see section 1.2.2, Converging films, equation 1.11).

It was indicated in section 1.2.1, Total closing force, that the difference in closing force was not the only reason why unbalanced seals are normally limited to 1 MPa g (145 lbf/in²) whilst balanced seals can be used up to 7 MPa g (1015 lb/in²) and above. It can be clearly seen that it is the difference in film formation that is important. Unbalanced seals seal progressively tighter as the pressure increases, but move more and more towards dry running, whilst balanced seals can potentially become more prone to leakage and instability as film formation improves.

Centrifugal effects

All dynamical mechanical seals have at least one rotating face and one stationary face. The fluid film between the faces must, therefore, see some rotation and be influenced by centrifugal effects.

In fact, the fluid adjacent to the rotating face rotates at the same speed, ω, as the face, whilst the fluid at the other side (the stationary side) is stationary (see Fig. 1.10, inset (a)). So despite its thinness, the fluid film contains a velocity profile, and centrifugal force will attempt to fling it out from between the faces to a degree that depends on the velocity in the particular radial plane.

The centrifugal force is opposed by surface tension and, in the normal configuration with the sealed fluid on the outer diameter of the sealing interface, by the pressure of the sealed fluid. However, in large seals it is possible to experience film starvation through centrifugal action, resulting in dry running, high heat generation, and short seal life.

Compressor seals fall into this category and it is wise to make a theoretical check that the barrier fluid/sealed fluid pressure differential is sufficiently high to guarantee proper face lubrication.

The calculations may be made as follows, using the nomenclature in Fig. 1.10.

Centrifugal film effects: theory

If

l_c = circumferential length of seal face element
m = mass of element in sector
r = radius of element
R_{ie} = inner radius of element
ω = speed of rotation of element
ϱ = density of fluid
$R_o = D_o/2$ and $R_i = D_i/2$

then

$$m = l_c \times \delta r \times \delta h \times \varrho$$

Force on ring element $= m \times r \times \omega^2$

With sector of unit length ($l_c = 1$)

$$\text{Force on sector} = \varrho \times \omega^2 \times \delta h \int_{R_{ie}}^{R_o} r\,\delta r$$
$$= \varrho \times \omega^2 \times \delta h[r^2/2]_{R_{ie}}^{R_o}$$

Force $= \varrho \times \omega^2 \times \delta h[R_o^2 - R_{ie}^2]/2$

This is balanced by a pressure force equal to $\Delta p_c \times 1 \times \delta h$

Thus

$$\Delta p_c \times \delta h = \varrho \times \omega^2 \times \delta h[R_o^2 - R_{ie}^2]/2$$

and

$$\Delta p_c = \varrho \times \omega^2 \times [R_o^2 - R_{ie}^2]/2 \qquad (1.14)$$

Example 7.

Consider a large seal running at 10 000 rev/min ($\omega = 1047$ rad/s)

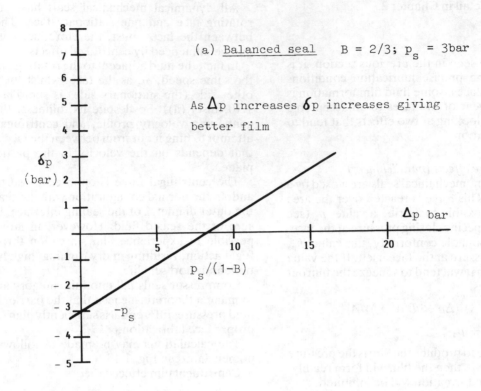

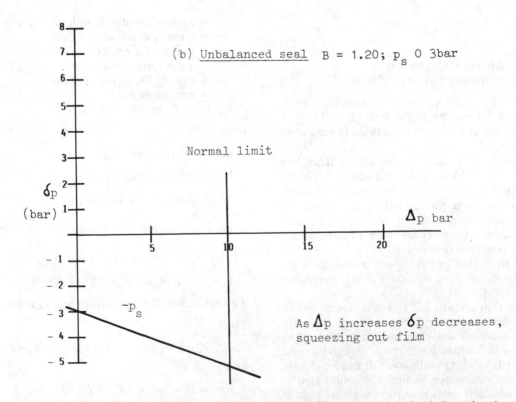

Fig. 1.9.   Influence of sealed pressure differential on *δp* in balanced and unbalanced seals

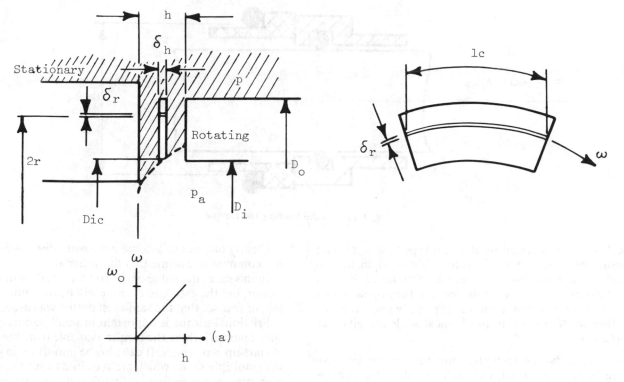

Fig. 1.10. **Centrifugal effects on the fluid film between rotating faces**

$D_o = 137.5$ mm and $D_i = 131.5$ mm: thus $R_o = 0.06875$ m and $R_i = 0.06575$ m

Assume a barrier fluid of oil with a specific gravity of 0.8 (density $= 0.8 \times 10^3$ kg/m^3).

We will calculate the necessary differential fluid pressure to ensure full film penetration of the face at the rotating face where the velocity is at a maximum ($\omega = \omega_o$).

$$\Delta p_c = 0.8 \times 10^3 \times 1047^2 \times [0.06875^2 - 0.06575^2]/2$$

$$= 1.77 \times 10^5 \text{ Pa}$$

$$= 0.177 \text{ MPa (25.7 lbf/in}^2\text{)}$$

If we calculate the required Δp_c on the basis of half rotational speed ($\omega = \omega_o/2$), ensuring full penetration at mid film thickness (a more generally accepted criterion) then

$$\Delta p_c = 0.177 \times (0.5)^2 = 0.044 \text{ MPa (6.4 lbf/in}^2\text{)}$$

For film formation, taking account of centrifugal effects

$$\Delta p > p_n + \Delta p_c$$

1.2.3 Friction, torque, and power

This section illustrates two approaches to the prediction of interface friction which are in common use. Other ways exist and, in appraising them, it is clearly important to see a validation of the method by correlation with measured results.

The first, empirical, method uses a general friction coefficient which relates to the total seal closing force, F_t, whilst the second method looks at viscous and dry friction coefficients from a more fundamental standpoint.

Frictional torque (Empirical method)

The frictional torque at the seal faces is a function of the closing force, the mean diameter of the seal faces, and the coefficient of friction, μ, between the faces. The calculation of frictional torque by this method is based on the use of a differential pressure, Δp, only and the assumption that p_a is negligible in the calculation of F_t (see equation 1.7a).

The hydraulic closing force is affected by the balance ratio, as already shown, but it is also affected by the face area (or face width). Wider faces create higher hydraulic closing forces and greater frictional torques (see note below).

Referring to Fig. 1.11

The tangential friction force at the face $= \mu \times F'_t$
where F'_t is defined as $F_s + \Delta p \times A_f \times B$
The face frictional torque,
$$M = \mu \times F'_t \times D_m/2$$
$$= \mu \times (F_s + \Delta p \times A_f \times B) \times D_m/2$$

The coefficient of friction in the above equation is an 'overall coefficient' in that it is applied to the total closing force of the seal. In specific cases the appropriate

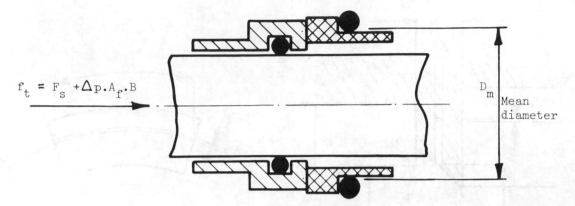

Fig. 1.11.   Axial loading and torque

coefficient will depend on the seal type, the operating conditions, and the fluid sealed. More sophisticated mechanical seals usually operate with lower friction coefficients than simple seals because they operate in a different lubrication regime. Typical values for the coefficient of friction in mechanical seals are given in Table 1.3.

Note. Wider faces do not give longer life simply because there is more material and, indeed, often fail prematurely because of the effects of higher heat generation. There are, however, circumstances where wider faces can give better life, but these usually relate to low pressure operation where high spring specific face pressures can preclude film formation and cause 'dry running' (see section 1.2.2, Film formation).

Example 8.

For the 70 mm balanced seal of Example 1, sealing water ($\mu = 0.1$) at 1 MPa g

$$F'_t = F_s + \Delta p \times A_f \times B$$
$$= 285 + 1 \times 920 \times 0.683 = 913.4 \text{ N}$$

Then frictional torque

$$M = 913.4 \times 0.0837/2 \times 0.1 \text{ Nm}$$
$$= 3.82 \text{ Nm}$$

It must be pointed out that the 'overall coefficient of friction' is not universally used; sometimes the frictional torque is calculated using a net closing force, the difference between the total closing force and the hydrostatic component of the opening force.

Clearly one has to assume a pressure distribution and it is common to assume that this is linear.

In this case, the value of the friction coefficient will be higher, but the estimated torque will be the same assuming, of course, that the same test datum was used.

Frictional torque is important in small pumps or high speed pumps, where the torque available from the driver at start-up is not high. It can also be important in pumps with multiple seals, which are required to start up under high pressure. Starting or 'break-out' torques may be several times higher than normal running values. For start-up it is advisable to use the dry coefficient of friction values given in Table 1.2. These values should be used in the selection of drivers and couplings.

Frictional torque (Basic principles)

Interface friction can be split into two components.

Viscous shear of the interface film, governed by Newton's Law of viscous shear which states that the friction force per unit area is proportional to the shear rate.

$$F \text{ (film)} = \eta \times V \times A_f/h \qquad (1.17)$$

where

η is the dynamic viscosity of the fluid
V is the relative sliding velocity
A_f is the face area
h is the effective film thickness

In practice, friction due to viscous shear is not easy to predict in a mechanical seal because the film thickness is not usually known. Similarly, the viscosity at the face is

Table 1.3.   Typical values of coefficients of friction for mechanical seals

Pressure (MPa g)	Water and aqueous solutions	Light hydrocarbons visc. <2 cSt	Light oils visc. 2–50 cSt	Heavy oils visc. >50 cSt
0 000	0.1	0.1	0.14	0.18
0.138	0.1	0.075	0.11	0.14
0.276	0.1	0.055	0.08	0.10
0.413	0.1	0.03	0.055	0.065
0.55	0.1	0.025	0.041	0.05
0.69	0.1	0.02	0.03	0.04
0.69+	0.1–0.04	0.02	0.03	0.04

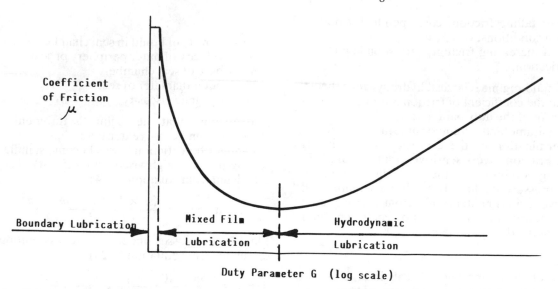

Fig. 1.12. Friction coefficient as a function of seal duty parameter (The minimum occurs at a G value between 10^{-8} and 10^{-7})

not precisely known since it depends on the face temperature.

Solid sliding contact, governed by Amonton's Law which states that friction force F is proportional to normal load.

$$F \text{ (solid)} = \mu \times F_m \tag{1.18}$$

where

μ is the friction coefficient

F_m is the mechanically supported load

Total friction force, $F = F \text{ (film)} + F \text{ (solid)}$ (1.19)

However, mechanical seals commonly, though not always, operate in an intermediate regime where there is some surface contact at asperity peaks, although a fluid film is also present. In these circumstances neither of the friction equations is applicable, unless the proportion of load carried by each mode is known, which it usually is not!

From section 1.2.2, Parallel films, it was shown that the load support may be apportioned if the pressure distribution factor β is known, which it seldom is.

The following example does, however, illustrate how estimates may be made using this method, having made suitable assumptions.

Example 9.

Following Example 3 with water as the sealed fluid and a speed of 3000 r/min we have the following

Mechanical load support, $F_m = 453.36$ N
Coefficient of friction (dry)
$\mu = 0.15$ (Table 1.2, tungsten carbide/carbon)
Viscosity, $\eta = 10^{-3}$ Ns/m^2
Face area, $A_f = 920$ mm^2
Mean relative sliding velocity

$$V = \frac{(80.17 + 87.17)}{2 \times 10^{-3}} \times \pi \times \frac{3000}{60} = 13.15 \text{ m/s}$$

Mean film thickness, $h = 0.5\,\mu$m (assumed as typical)
then

$$F \text{ (solid)} = \mu \times F_m = 0.15 \times 453.36 = 68.0 \text{ N}$$
and

$$F \text{ (film)} = \eta \times V \times A_f/h$$

$$= 10^{-3} \times 13.15 \times \frac{920 \times 10^{-6}}{0.5 \times 10^{-6}} = 24.2 \text{ N}$$

Total friction $F = 68.0 + 24.2 = 92.2$ N

Total torque $M = F \times D_m/2 = 92.2$

$$\times \frac{(80.17 + 87.17)}{2 \times 10^{-3}} \times \frac{1}{2} = 3.85 \text{ Nm}$$

This value agrees with the value of 3.82 Nm given earlier in this section.

Duty parameter

A non-dimensional parameter, analogous to the Stribeck Number in plain bearings, can be used to show the relationship of the coefficient of friction to the lubrication mode in the seal. The duty parameter, G, for a seal is given by

$$G = \eta \times V \times b/F_t \tag{1.20}$$

where

V = mean face velocity
η = dynamic viscosity
b = face width

The severity of the contact conditions decreases with increasing G.

Figure 1.12 shows the relationship between the coefficient of friction and the Duty Parameter. This distinguishes the three distinct regimes of lubrication:

a high friction regime at low G, corresponding to boundary lubrication;

a region of falling friction, corresponding to mixed lubrication conditions;
a region of increasing friction, corresponding to full fluid lubrication.

For the full fluid regime, classical hydrodynamic theory predicts that the coefficient of friction is proportional to the square root of the duty parameter.

The duty parameter has limited value in application as far as determination of the coefficient of friction is concerned. Friction is very sensitive to small changes in the operating conditions as the faces distort through thermal or pressure loading. Instead of a smooth, well defined curve, experimental results from seals exhibit a very large amount of scatter in friction coefficient for a given value of the duty parameter.

Absorbed power

Power is absorbed by mechanical seals in at least two ways. The dominant component under normal operating conditions originates in face friction. Fluid losses within the seal chamber can also absorb substantial amounts of power, particularly in high speed applications (over 5000 r/min) on larger seals (over 100 mm).

Face friction
The previous section was devoted to predicting the frictional torque at the seal face. The power absorbed by a seal, H_s, is the product of that torque times the angular velocity.

$$H_s = M \times \omega$$
$$= M \times 2\pi \times n \qquad (1.21)$$

where

n is rotational speed
M is frictional torque

Thus in the case of the overall coefficient of friction

$$H_s = \mu(F_s + \Delta p \times A_f \times B) \times D_m/2 \times 2\pi \times n \qquad (1.22)$$

Example 10.

In Example 8, a frictional torque of 3.82 Nm was calculated for the 70 mm seal. For a speed of 3000 r/min, the absorbed power is

$$H_s = 3.82 \times 3000 \times 2\pi/60$$
$$= 1200 \text{ W}$$

Power losses in general purpose seals are seldom higher than a few kilowatts, although in the case of large high speed seals in compressors, or on high pressure pipe line seals they can be as high as 20 kW.

Seal chamber fluid friction losses
Fluid friction losses in the seal chamber can be significant under both laminar and turbulent fluid conditions. Different formulae are used, depending on the flow regime. The relevant regime is established by determining the rotational Reynolds Number, *Re*.

$$Re = \varrho \times V_s(d_{sc} - d_s)/(2 \times \eta)$$

where

ϱ = density of liquid in seal chamber
V_s = velocity of outer periphery of seal
d_{sc} = bore of seal chamber
d_s = outer diameter of seal
η = dynamic viscosity

It is assumed that the regime is turbulent when the Reynolds Number is greater than 2000.

Laminar flow friction losses become significant when sealing fluids of high viscosity (>25 cSt). They can be calculated from equation (1.24)

$$H_{sc} \text{ (laminar)} = \frac{\{\pi \times l_r \times \eta \times \omega^2 \times d_s^3\}}{\{2 \times (d_{sc} - d_s)\}}$$

Fluid friction losses under turbulent conditions can be calculated from equation (1.25)

$$H_{sc} = \frac{1.53 \times 10^{-2} \times \varrho \times \omega^3 \times d_s^4 \times l_r}{Re^{0.43}} \qquad (1.25)$$

l_r = length of rotating component

Example 11a.

Consider the 70 mm balanced seal operating on oil at 7000 r/min

Angular velocity ($= 2 \times \pi \times 7000/60$), $\omega = 733$ rad/s
Take seal outer dia = 95 mm, $d_s = 0.095$ m
Seal chamber diameter = 100 mm, $d_{sc} = 0.10$ m
Length of rotating component = 75 mm, $l_r = 0.075$ m
Density of oil, $\varrho = 800$ kg/m^3
Viscosity of oil (50 cP), $\eta = 50 \times 10^{-3}$ N·s/m^2

Reynolds number, *Re*

$$= \frac{800 \times \pi \times 0.095 \times (7000/60) \times (0.1 - 0.095)}{2 \times 50 \times 10^{-3}}$$

$$= 1393 \quad \text{and is } <2000, \text{ so flow is viscous}$$

Churning loss H_{sc} (laminar)

$$= \frac{\pi \times 0.075 \times 50 \times 10^{-3} \times 733^2 \times 0.1^3}{2 \times (0.1 - 0.095)}$$

$$= 633 \text{ W}$$

Example 11b.

Consider the same 70 mm balanced seal operating on water

Density of water $\varrho = 1000$ kg/m^3
Viscosity of water (1 cP) $\eta = 1 \times 10^{-3}$ N·s/m^2

Reynolds number, *Re*

$$= \frac{1000 \times \pi \times 0.095 \times (7000/60) \times (0.1 - 0.095)}{2 \times 1 \times 10^{-3}}$$

$$= 87048 \quad \text{and is } >2000, \text{ so turbulence exists}$$

Churning loss, H_{sc}

$$= \frac{1.53 \times 10^{-2} \times 10^3 \times 733^3 \times 0.095^4 \times 0.075}{87048^{0.43}}$$

$$= 277\ \text{W}$$

The effects of different viscosities, etc., may readily be calculated.

A reasonable estimate of the absorbed power in a mechanical seal may be made by combining the face friction and fluid friction losses

$$H = H_s + H_{sc} \tag{1.26}$$

There are a number of empirical methods available which also give reasonable estimates of absorbed power in a simpler way.

The following formula is useful when exact seal details are not known. It has been validated by comparison with results from research papers and seal vendors catalogues.

$$H'_s = N \times \mu\{(2.297 \times 10^{-8} \times \Phi^{2.215} \times P) + (1.216 \times 10^{-7} \times \Phi^{2.17})\} \tag{1.27}$$

where

H'_s = heat generated (kW)
Φ = nominal seal size (mm)
P = pressure (bar)
N = speed (r/min)

Example 12.

In the case of the 70 mm seal, assume we are sealing water at 1 MPa g speed 3000 r/min.

$$\mu = 0.1$$

and

$$H'_s = 3000 \times 0.1\{(2.297 \times 10^{-8} \times 70^{2.215} \times 10) + (1.216 \times 10^{-7} \times 70^{2.17})\}$$
$$= 1.21\ \text{kW}$$

This compares very well with the 1.20 kW calculated in Example 10 and with experimental results.

Absorbed power is a very important factor in seals as will be discussed in the next sections.

1.2.4 Cooling and circulation requirements

The heat generated by a mechanical seal must be dissipated to prevent temperatures in the vicinity of the faces rising to an unacceptable level. With a dead-ended seal chamber (viz no flow through the seal chamber) a thermal balance is established with the heat being conducted into the surrounding liquid and thence to the pump casing. For more severe condition (larger seals, higher speeds and pressures) it is necessary to provide circulation flow through the seal chamber to remove heat. Figure 1.1 shows a simple arrangement where flow is provided from the pump discharge through the seal chamber to the back of the pump impeller. It is normal

to 'size' this flow on the basis of an allowable temperature rise.

If the power dissipated in the seal is H, the coolant flow Q required is

$$Q = H/(\varrho \times c \times \delta T_c) \tag{1.28}$$

where

ϱ = coolant density
c = coolant specific heat
δT_c = coolant temperature rise

The choice of the allowable temperature rise is somewhat arbitrary; it is set at a low value in order to sensibly limit the temperature gradients within the seal components themselves and thus limit the chance of undesirable thermal distortions and stresses. A typical figure used in design is 25 K maximum.

Example 13.

Taking the calculated power loss of 1200 watts from Example 10 for a 70 mm seal operating on water at 3000 r/min, the following minimum value of cooling flow rate is obtained

ϱ = 1000 kg/m^3
c = 4200 J/kg·K

$$Q = \frac{6 \times 10^4 \times 1200}{1000 \times 4200 \times 25} = 0.7\ \text{l/min.}$$

However, a practical minimum flow rate of 10 l/min will usually be chosen. This avoids the problems associated with small orifices for flow control and provides more than adequate cooling in most applications. A more careful assessment is only required for large seals (>75 mm) operating at higher speeds (>3000 r/min) and/or pressures (>30 bar).

1.2.5 Seal face deflections

Flatness and parallelism of the sealing faces are fundamental to the effective performance of a mechanical seal. Experience shows that the faces should be initially lapped to a flatness of about 0.001 mm (3 to 4 helium light bands), maximum peak-to-valley height; this is normally provided by the manufacturer. It follows that disturbances of the flatness of this order of magnitude can affect performance adversely and should be avoided. Sources of such disturbances are associated with the casing stiffness and tolerances of the pump, as well as the seal and its operating conditions.

Two types of seal face disturbance are particularly likely; they correspond to the two most compliant modes of elastic deflection of a ring.

(1) Axisymmetric coning. This radial taper causes the interface profile to be divergent or convergent and has a powerful effect on hydrostatic pressure in the interface. It is caused by various moments acting about the neutral axis through the seal's cross-section centroid. The moments in question are uniformly distributed around the circumference of the seal and therefore produce

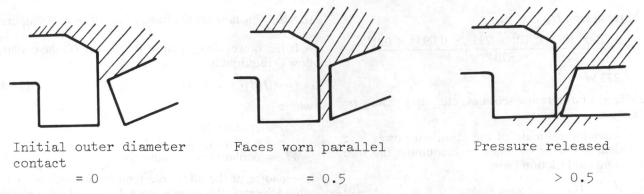

Initial outer diameter   Faces worn parallel   Pressure released
contact
          = 0   = 0.5   > 0.5

Fig. 1.13.   Leakage after running under pressure with concave deflections

axisymmetric coning. It is possible to ascertain the degree of coning by calculation; often this is done by applying ring theory, since seal rings are relatively short, but finite element analysis can also be used, although particular care is necessary to ensure that moments are accurately reproduced. Some special designs intentionally incorporate a convergent profile in the direction of leakage flow in order to ensure a thick fluid film with hydrostatic pressure carrying the closing force, as in a hydrostatic bearing. The leakage of such a seal is normally high, but is traded against the improved reliability expected from total separation of the sliding faces.

(2) Waviness. This is commonly found as two cycles round the circumference of the seal face and can usually be seen in used seals. It may have a powerful effect on hydrodynamic pressure in the interface. It can be caused by non-uniform diametral loads, such as drive-lug forces or non-uniform mounting reactions. Other sources can be residual stress and instability of constructional materials. In large-diameter seals it is particularly important that such effects are minimized during manufacture. Calculation of waviness is not practicable.

Coning can arise from hydraulic, mechanical, or thermal effects, or any combination of them.

Hydraulic deflections

Hydraulic deflections are caused by various fluid pressure loads distributed over the seal surfaces, both radially and axially directed loads have to be considered. The axial loads include a large resultant reaction-force at the sealing interface. The line of action of this varies depending upon details of coning and fluid pressure distribution in the interface.

Mechanical deflections

Mechanical deflections can arise from such sources as spring load, and the interference forces and friction forces of secondary sealing rings, such as the 'O' ring at the balance diameter. The latter become increasingly important in large seals.

Structural deflection of the pump casing can have a serious effect on seal performance. This can arise from internal effects, such as hydraulic or thermal distortions,

and from external effects such as flange loadings by pipe work. Shaft stiffness and bearing details are also important for critical duties. A stiff pump can be expected to give significantly better seal reliability than a lower-cost less stiff design (see Chapter 6, section 6.2.4).

Thermal deflections

Thermal deflections result from a non-uniform temperature distribution within the seal. Frictional heat, conducted through the body of the seal from the sliding interface to the seal-chamber fluid, is usually the main factor, but when sealing a high-temperature fluid additional thermal gradients occur. The direct effect of thermal expansion is a major component of the thermal deflection, but thermal stresses can induce an additional deflecting moment which must also be considered.

Calculation of thermal deflections is possible if reliable data are available for frictional heat generation, but this depends upon the face materials and the operating conditions. It is not sufficient to assume a universal value of friction coefficient. Definition of values of heat transfer coefficient, between seal and contacting fluids, can also present difficulty.

Wear

Provided the deflections are not so great that the initial leakage is unacceptable, the faces tend to bed together by running-in. This can be a slow process with hard face combinations, with less initial distortion permissible if satisfactory running conditions are to be achieved. Even with a soft face excessive bedding-in to accommodate distortions has to be avoided; otherwise there will be a risk of leakage when the bedded-in seal is stationary. This can also be a problem in seals subject to considerable pressure variations in service, e.g., with submarine seals.

This problem of leakage at low pressure with bedded-in seals is illustrated in Fig. 1.13. In the running-in stage, with the seal ring distorted concave at high pressure, the faces contact over a small area at the outer periphery. This means that the average fluid pressure across the faces will be low ($\beta < 0.5$) and therefore the residual load support will be high, giving a tight seal.

Table 1.4. Approximate *PV* limits in bar m/s for general seals with various combinations of seal face materials and fluids

Face material combination	Water and aqueous liquids		Other liquids	
	Unbalanced	Balanced	Unbalanced	Balanced
Carbon				
Stainless steel	5	—	30	—
Lead bronze	25	—	35	—
Stellite	25	85	50	580
Alumina	35	210	90	420
Chrome oxide	70	420	—	—
Tungsten carbide	70	420	90	1220
Silicon carbide	90	630	90	1850
Tungsten carbide				
Tungsten carbide	45	260	70	420
Silicon carbide	60	360	90	1050

After a period of running, the faces will bed-in to be parallel across the whole face area whilst the seal is under high pressure. In this condition the seal performs well. However, if the pressure on the seal is reduced, then the distortion of the carbon face will also reduce and what was a pair of parallel faces at high pressure now becomes a pair of convergent faces. Convergent faces produce high average fluid film pressures ($\beta > 0.5$), which can push the faces apart and produce leakage (see section 1.2.2, Converging films).

1.3 OPERATING LIMITS

While there are many factors which govern the selection of, and limitations on, mechanical seals, two principle ones are the limit of the seal face as a loaded bearing and the limit of fluid film stability.

1.3.1 Load bearing capacity (*PV* − limit)

As indicated in section 1.2.2, the face load in most stable mechanical seals is supported both mechanically and by the fluid film. The mechanical part of the load is supported by contact between the high spots on the seal faces and, hence, most seals will wear and have a finite life, determined by the seal face loading, the relative velocities of the faces and the behaviour of the face materials. In some cases the limitation is not wear rate but the breakdown of the seal face as a load supporting bearing as its capacity is exceeded. This can result in phenomena such as 'thermal cracking' of hard faces.

A *PV* factor is commonly used to express the severity of face contact conditions and the allowable operating limit of seals to ensure adequate life, where *P* is some measure of the face load (usually the pressure of the sealed fluid) and *V* the mean peripheral velocity of the seal faces. It has to be appreciated that *PV* is no more than an approximate guide to seal performance. It is influenced not only by the face materials and the lubricating properties of the sealed fluid, but is also very dependent on the seal type. The seal type affects the

apportionment of load support and, since *PV* is a measure of the dry-running properties, it is more relevant to seals where a high proportion of the load is carried by asperity contact.

PV does not take into account the effect of such external factors as misalignment, vibration, and the presence of abrasive solids in the sealed fluid or the lubrication regime. At the limits of *P* and *V*, the simple relationship between *PV* and life does not hold, and in such cases it is better to present the data graphically in terms of the separate variables.

Definition of PV factors

The *PV* factors for materials and seals respectively are defined in (i) and (ii) below. Notice the difference in units which reflect differences in historical usage.

(i) The *pV* factor for a sliding material combination is defined as follows.

p = the contact pressure in MPa
V = the mean sliding-velocity at the sealing interface in m/s

hence '*PV* factor' = $p \cdot V$ (MPa · m/s)
This has been widely used in bearing technology.

(ii) The *PV* factor for a mechanical seal is commonly defined as follows.

P = the pressure drop across the seal in bar
V = the mean sliding-velocity at the sealing interface in m/s

i.e.

$$V = \pi\{(d_o + d_i)/2000\} \times N/60$$

with d_o and d_i in mm and N in r/min

Hence '*PV* factor' = $P \times (d_o + d_i) \times N/38200$

(bar m/s)   (1.29)

Variations of this definition for mechanical seals have been used but, in the interests of standardization, the above is strongly recommended. Until a standard form is in common use, it is necessary to ensure that the actual definition used for *PV* is quoted.

Some sources compute *PV* on the basis of a net mechanical face pressure, having first subtracted the hydrostatic support load, and some use the balance diameter rather than the mean face diameter.

Additionally, some experimental data are based on mechanical loading (by means of a spring lever) only, whilst other data are based on actual seal tests. Quoted values can thus vary significantly. The values shown in Table 1.4 are based on seal test data. Higher values can be obtained with special designs.

Example 14.

For the 70 mm seal operating at 3000 r/min with a pressure differential of 10 bar

$$PV = 10 \times (80.17 + 87.17) \times 3000/38200$$
$$= 131 \text{ bar m/s}$$

1.3.2 Heat transfer (ΔT limit)

The majority of mechanical seals operate with at least a partial fluid film between the faces with the possible exception of unbalanced seals at high pressures where dry friction dominates and *PV* is most meaningful.

The frictional heat generated at the face of a mechanical seal can lower the viscosity of the fluid in the film to the point where the load bearing ability of the film is insufficient and failure results through heavy wear or face damage. It can also heat the film to such an extent that the fluid boils or vaporizes at the prevailing film pressure; seal face vaporization can often be seen or heard as an intermittent banging or popping sound and it results in severe face damage and eventual heavy leakage.

It is a principal objective in design and selection to ensure that a stable lubricating fluid film is both established and maintained. Thus, the fluid film temperature is all important; various methods are available for calculating it. The following method is empirical in the sense that it calculates a single face temperature and compares it with the boiling point (or bubble point for liquid mixtures) of the sealed fluid at the sealed pressure. (More complex analysis (e.g., finite element) can identify temperature variations in the film and compare them with the boiling point at that point in the film; this depends on an accurate knowledge of the pressure distribution.)

The heat balance equations (see Fig. 1.14) contain friction and heat transfer values which agree closely with experimental values and the method is reliable when applied to general purpose seals. Its principal value here, however, is to serve to illustrate those factors which influence seal face temperatures so that choices may be made.

Relating to Fig. 1.14

(1) Heat generated at the faces

$$H_g = \mu D_m n (F_s + \Delta p \times A_f \times B) \qquad (1.30)$$

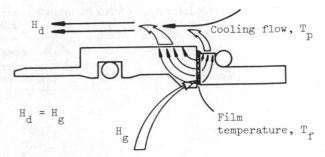

Fig. 1.14.   The heat balance equations for a mechanical seal

(2) Heat dissipated

$$H_d = m'kA \tanh(m'l) \times \Delta T$$

where

$m' = \sqrt{(hC/kA)}$
k = thermal conductivity of seal
A = cross sectional area perpendicular to heat flow
h' = heat transfer coefficient
l = axial length of seal ring heat transfer surface
C = circumference of heat transfer
ΔT = temperature rise at seal face over surrounding fluid ($T_f - T_p$)

Heat is dissipated from both rotating and stationary faces and hence

$$H_d \text{ (total)} = H_d \text{ (rotating face)} + H_d \text{ (stationary face)}$$
$$= \Delta T[\{m'kA \tanh (m'l)\}_1$$
$$+ \{m'kA \tanh (m'l)\}_2]$$

By equating heat generated and heat dissipated, the temperature rise (ΔT) at the faces may be calculated.

Typical values of constants are as follows.

(1) Thermal conductivity (W/mK)

Plain carbon	5.2
Metallized carbon	15
18/8 S.S.	15
16/2 S.S.	20
Cast iron	43/52
Bronze	43/52
Copper alloy	204
Copper	395
Tungsten carbide	87
Silicon carbide (Refel)	133
Alumina	3

(2) Heat transfer coefficients (W/m²K)

	Water and aqueous solutions	Light hydrocarbons <2 cSt	Light oils 2–50 cSt	Heavy oils >50 cSt
Rotary seal	5700	4300	2000	1250
Stationary seal	5700	2000	800	570

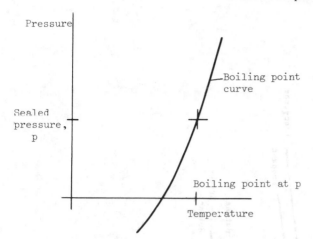

Fig. 1.15.   Typical boiling point curve for sealed liquids

An estimate may thus be made for the face temperature and the influence of the materials of construction may be seen from the equations and tables.

If a fluid is being sealed at a particular pressure then it has a boiling point according to its boiling point curve (Fig. 1.15).

The calculated value of ΔT to prevent face vaporization thus represents the temperature rise at the face and the maximum operating temperature of the seal is ΔT degrees below the fluid boiling (or bubble) point at the sealed pressure. (In fact, because there is a pressure drop across the faces, vaporization will take place at a pressure lower than the sealed pressure. The ΔT required will then relate to the boiling point at that lower pressure.)

By considering various pressures, an operating curve may be drawn; it will be bounded at the top by the *PV* limit (Fig. 1.16).

The value of ΔT is often referred to as the 'ΔT required'.

The difference between the boiling point of the pumped fluid at the sealed pressure and the pumping temperature is referred to as the ΔT (available).

For stable operation

'ΔT available' $>$ 'ΔT required'

Note. It is not uncommon to use the alternative concept of Δp to define the margin to fluid film boiling. Indeed API 610 has historically proposed a Δp margin of 25 lb/in^2 (0.17 MPa) for all marginal duties regardless of sealed fluid, speed of rotation, seal size, sealed pressure, etc.

It should be noted that the gradient of the vapour pressure curve depends very much on the fluid being pumped and indeed the actual pressure value on the vapour pressure curve. Figure 1.17 gives vapour pressure curves for a range of light hydrocarbons. From this it can be seen that for a fixed pressure margin of 0.17 MPa the ΔT for ethylene decreases from 4°C at 0.5 MPa to 1°C at 2.5 MPa, that for *n*-pentane from 14°C at 0.5 MPa to 9°C at 2.5 MPa. Thus the use of Δp as a concept becomes dangerous on liquids with a steeply sloping vapour pressure curve.

1.3.3   Low speed limit

In addition to the *PV* limit that operates at high speeds, there is some evidence to suggest that there is also a low speed limit when there is a transition from fluid-film to boundary conditions. This corresponds to a low value of G as discussed in section 1.2.3 on Duty parameter. Although seals can operate under boundary conditions, this will be associated with face wear and it may be preferable to adjust the conditions in the seal to ensure fluid film lubrication.

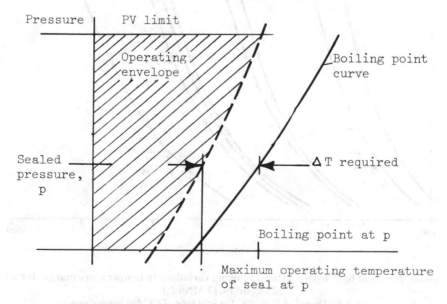

Fig. 1.16.   Pressure/temperature operating envelope for a seal

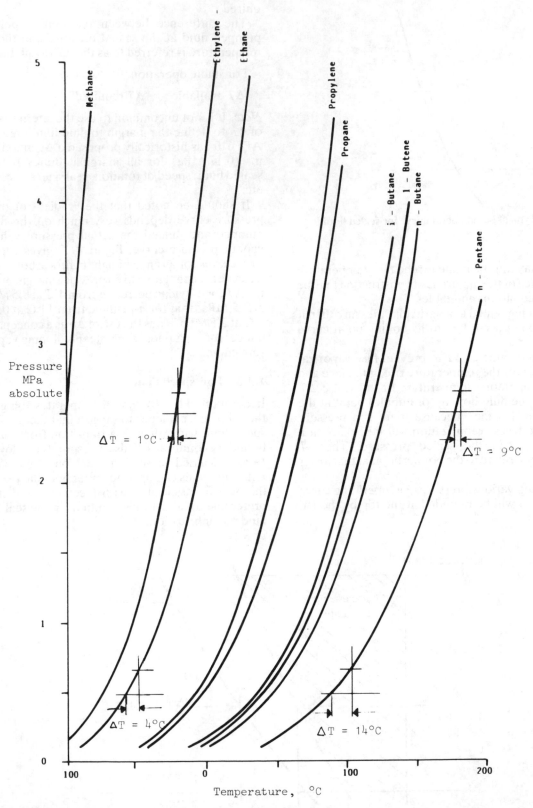

Fig. 1.17. Vapour pressure curves of light hydrocarbons showing variations in temperature margin for a fixed pressure margin of
25 lb/in² (0.17 MN/m²)
At 0.5 MN/m², ΔT = 4°C for ethylene, 14°C for n-pentane
At 2.5 MN/m², ΔT = 1°C for ethylene, 9°C for n-pentane

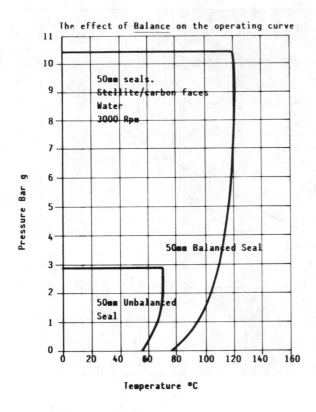

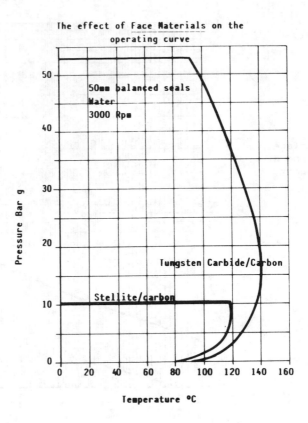

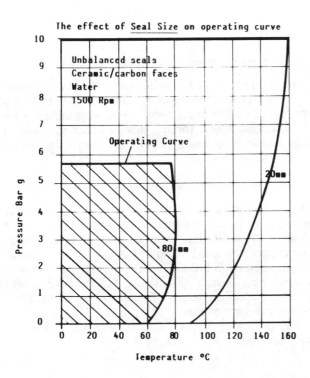

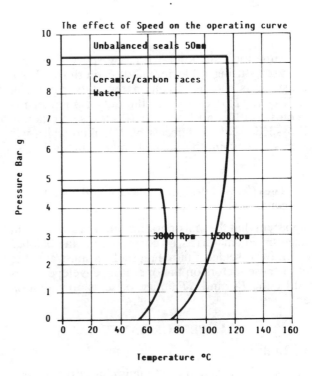

Fig. 1.18. The effect of balance, face material, seal size, and operating conditions on performance

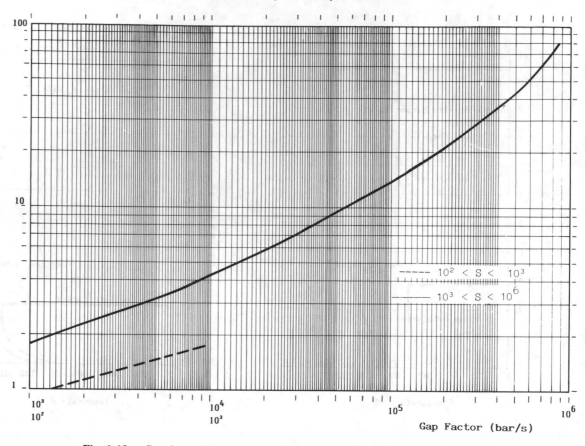

Fig. 1.19.   Gap factor *S* for use in seal leakage calculations

This situation can occur with low speed autoclave agitator seals using a barrier fluid. Experimental evidence suggests the minimum in the duty parameter curve (Fig. 1.12) occurs when the duty parameter lies between 1×10^{-8} and 1×10^{-7}. It is a wise precaution for low-speed seals to select a barrier fluid which will have a viscosity in the seal chamber that will give $G > 5 \times 10^{-8}$.

1.3.4   The effects of design, materials, and operating conditions

It is apparent that the seal operating limits are dependent on size, speed, materials of construction, the balance ratio, and the sealed fluid. Figure 1.18 illustrates the effect of these factors on the operating envelope determined by the PV and ΔT limits for seals operating on water.

1.4   PERFORMANCE

1.4.1   Leakage

All mechanical seals that are functioning correctly leak to some extent. The leakage rate depends on a number of factors. Not least of these is the lubrication regime in which the seal is operating.

Leakage under boundary lubrication conditions

Under conditions of full solid contact with no fluid film present, leakage is virtually nil. Operation in the boundary regime is, however, normally unacceptable because of high wear rates, though it may occur in some unbalanced seals operating near their pressure limit (about 10 bar g).

Leakage under mixed lubrication conditions

The majority of mechanical seals operate in a mixed lubrication regime, where the load support is by a combination of fluid film pressure and asperity contact. Leakage is by a combination of simple viscous flow and a pumping action induced by the relative motion of the asperities on the mating seal rings. Since asperity contact occurs, the leakage gap is a function of the surface finish of the seal components.

The following empirical equation for leakage has been proposed by Mayer (1977)

$$Q_1 = 3.6\pi \times d_o \times P \times h_s^2 \times S/P_g^2 \quad \text{(ml/h)} \quad (1.31)$$

where

d_o = entry diameter to the seal interface (mm)
P = pressure drop across seal (bar)

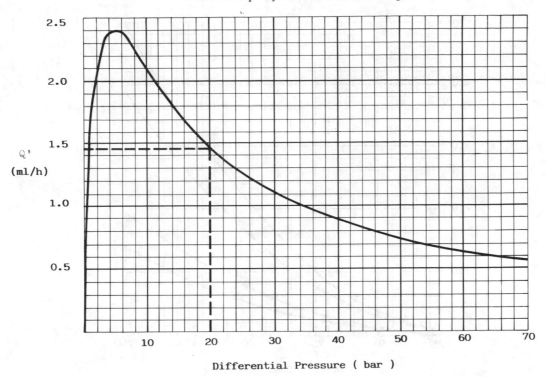

Fig. 1.20. Empirical leakage curve for 75 mm balanced seal operating at 3600 r/min

h_s = face separation (mm)
(typically 0.38×10^{-3} for bedded-in seals)
S = gap factor (see Fig. 1.19) (bar/s)
P_g = seal face pressure (bar)
= $P_s + B \times P$ (P_s in bar)

The face separation h_s is empirical and based on surface finish considerations and is not strictly the same as the interface film thickness, h.

This expression and the associated work carried out during its derivation, have been well supported by field experience on a wide variety of seals. It will be seen that leakage is dependent on seal size, balance, and the pressure drop.

Example 15.

Calculate the leakage from the 70 mm seal operating at 3000 r/min with a sealed pressure of 20 bar g.

d_o = 87.17 mm
d_i = 80.17 mm
P = 20 bar
h_s = 0.00038 mm
B = 0.683
P_s = 3 bar
P_g = $0.683 \times 20 + 3$ = 16.7 bar
S = 9×10^4 (from Fig. 1.19 with face velocity
= 13.1 m/s)
Q_l = $3.6 \times \pi \times 87.17 \times 20$
$\times 0.00038^2 \times 9 \times 10^4/16.7^2$
= 0.92 ml/h

An alternative graphical method, which depends on an empirical leakage rate, Q', for a 75 mm balanced seal operating at 3600 r/min (Fig. 1.20), together with a correction factor, K, for other sizes and speeds (Fig. 1.21) allows prediction of the leakage without a detailed knowledge of the seal parameters.

$$Q_l = Q' \times K \tag{1.32}$$

Referring to Figs. 1.20 and 1.21.

(1) The shape of the leakage vs pressure curve in Fig. 1.20 is characteristic of the mechanism of fluid film formation. The peak, which occurs at approximately 4.5 bar, corresponds to a condition where the effects of spring load, hydraulic closing force and film pressure combine to produce a maximum film thickness.

(2) The increase of leakage rate with the speed of rotation, illustrated in Fig. 1.21, demonstrates that the leakage is not simply a function of viscous flow between parallel boundaries; in addition, there is a significant pumping action by the relative motion of the asperities on the seal faces.

(3) A number of researchers have found that leakage of seals operating in the mixed lubrication regime is not directly dependent on the viscosity of the fluid (as it would be in the case of a fully hydrodynamic seal for example). This conclusion is borne out by experience in the field.

(4) It follows from (2) above that the leakage rate will be dependent on the surface roughness of the sealing faces. Although this can be detectable when testing new seals under laboratory conditions, no

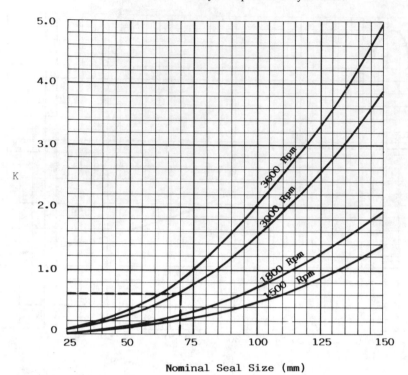

Fig. 1.21.   Seal leakage correction factor, *K*

significant difference can normally be observed in practice.

(5) Newly installed seals will sometimes leak more heavily than predicted, but the leakage rate then reduces after a period of time as the seal faces run in to compensate for distortions. Obviously, this wearing process will not take place as rapidly on seals having two hard faces (e.g., tungsten carbide vs tungsten carbide), so a higher degree of variability may be expected with seals using such face materials.

(6) In the case of volatile products, the visible leakage will be much less than predicted, due to evaporation. The extent to which this occurs is governed not only by the nature of the fluid, but by the operating temperature of the pump and the amount of frictional heat generated at the seal faces. In general terms it has been found, however, that actual leakage rates are much the same.

Example 16.

The above method gives the following leakage for the 70 mm seal of Example 15.

$$Q_1 = Q' \times K$$
$$Q' = 1.45 \text{ ml/h} \quad \text{(from Fig. 1.20)}$$
$$K = 0.6 \quad \text{(from Fig. 1.21)}$$

Thus, expected leakage rate

$$Q_1 = 1.45 \times 0.6 = 0.87 \text{ ml/h}$$

The leakages calculated by this method are average

values and, in practice, there may be quite large variations in performance between apparently identical seals operating on the same duty conditions. The following factors can be used to give confidence levels for the maximum leakage rate.

Confidence level	Factor
95%	×2
99%	×10
100%	×50

For example, if the calculated average leakage rate is 0.87 ml/h, then 95 per cent of all seals operating on that duty can be expected to leak at a rate less than 1.74 ml/hr, whilst 99 per cent will leak less than 8.7 ml/h.

Acceptable leakage limits

It is sometimes asked at what level of leakage a seal should be considered to have failed. Although the answer to this question will depend very much on the characteristics of the sealed product and the environment in which the seal is operating, as a general rule it may be said that a seal whose leakage rate is 250 times the theoretical average figure is definitely functioning incorrectly.

Leakage with full film lubrication

The problem of predicting leakage rate in seals with full film lubrication is complex. The equations for flow in thin films of specified shape are well-known, but the film shape and thickness in the sealing interface of a mechanical seal are not easy to define because of surface waviness, thermal and hydraulic distortions that depend on the particular design, and materials of construction, as well as operational factors such as pressure and speed.

A number of equations have been put forward for the calculation of leakage through seals operating in the full fluid lubrication regime. These are all based on the Poiseuille equation for laminar flow through a simple annular gap.

$$Q_1 = \frac{\pi D_m h^3 \Delta p}{12 \eta b}$$

None have, however, been validated by results from mechanical seals. In view of the doubts in their applicability and their extreme sensitivity to the value chosen for the interface film thickness, none can be recommended. Suffice to say, they will give leakage rates at least three orders of magnitude greater than that for the mixed lubrication case. It should be noted that, apart from exceptional cases, this high leakage militates against seals being designed for full fluid film lubrication.

1.4.2 Life

Seals rarely wear out through old age: other factors almost always precipitate premature failure. Because of this the life of a seal must be regarded as a statistical parameter rather than precisely definable. The effective working life of a seal is very dependent on details of the application. It will be increased by attention to the following.

Creation of a favorable working environment in the seal chamber:
    effective temperature control;
    exclusion of abrasive solids;
    development of an effective film between the faces (using a double-seal and barrier fluid if necessary).

Adequate technical specification of the seal.

Use of a high-grade pump having stiff housing and shaft, and high-quality bearing system.

It is important to take account not only of the first-cost of the seal (and pump), but also of the costs associated with failure, these include the following.

Lost production.

Cost of stand-by equipment.

Damage caused by escaping fluid:
    to personnel;
    to environment;
    to equipment (including bearings adjacent to seal).

Labour and parts for seal replacement, including spares stock.

Extremes of life that have been reported range from a few days on aggressive chemical duties to over 20 years of satisfactory service in favourable light hydrocarbon applications.

1.4.3 Reliability

In addition to the wide range of life in different duties mentioned in the previous section, seal life also varies appreciably for apparently identical seals in the same application, a situation similar to that experienced with rolling bearings. Because of this, it is appropriate to apply statistical concepts in discussing seal reliability.

The following two statistics have been applied to mechanical seals.

L_{10} Life.
This is defined as the running time at which the number of failures from a sample population of components has reached 10 per cent. Its use is particularly associated with Weibull failure analysis methods.

Mean time between failures (MTBF)
This is the arithmetic mean of the time between successive failures in a population.

$$\text{MTFB} = \frac{L_1 + L_2 + \cdots L_n}{n} \qquad (1.33)$$

where $L_1,, L_2$, etc., are the lives of the individual seals.

(*Note.* 'Population' here is used in the statistical sense of a number of nominally identical components subjected to identical operating conditions.)

Weibull analysis is of value in failure analysis as it enables the temporal distribution of the failures to be determined. [The technique of Weibull analysis is described in Carter (1972).] The Weibull Index β', which is obtained from the analysis, has the following values for different types of distribution:

early life failures (infant mortalities) $\beta' = 0.5$
mid-life failures (random failures) $\beta' = 1.0$
wear out failures $\beta' > 3.4$

Figure 1.22 shows a Weibull plot of failures of mechanical seals from a population of 29 methanol pumps. The value of 0.8 indicates a mixture of early life and random failures. It is typical of mechanical seals that failures follow an early life and random pattern rather than wear out. The L_{10} life, which can be read directly from the Weibull plot, is 46 days (1100 hours). Note that the analysis included data from survivors (i.e., seals that were still in service without failure, involving some that were still operating satisfactorily after 5 years – 40 000 hours).

The L_{10} life is very low compared to the value of 25 000 minimum that would be expected from rolling element bearings in a process pump. The low value implies a very large scatter in lives and with such distributions MTBF is a more meaningful statistic to use than L_{10} life.

The value of statistical analyses of the Weibull and MTBF types is that they allow comparisons and are particularly useful in determining whether design or operational changes have had the desired effect in improving life.

Figure 1.23 shows Weibull plots for seal failures from nitric acid circulation pumps, a difficult application where seal performance was particularly poor. Note

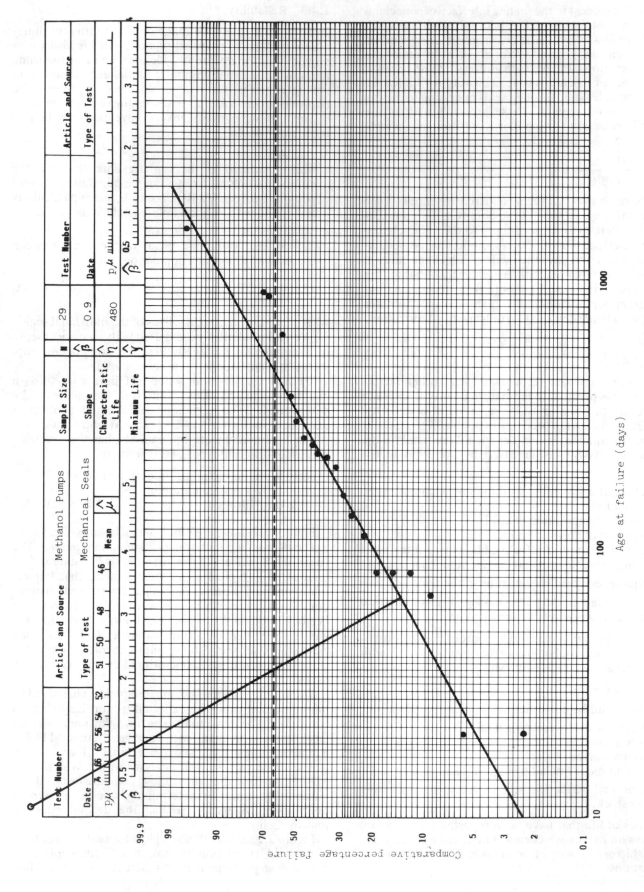

Fig. 1.22.   Weibull plot of seal failures from population of methanol pumps

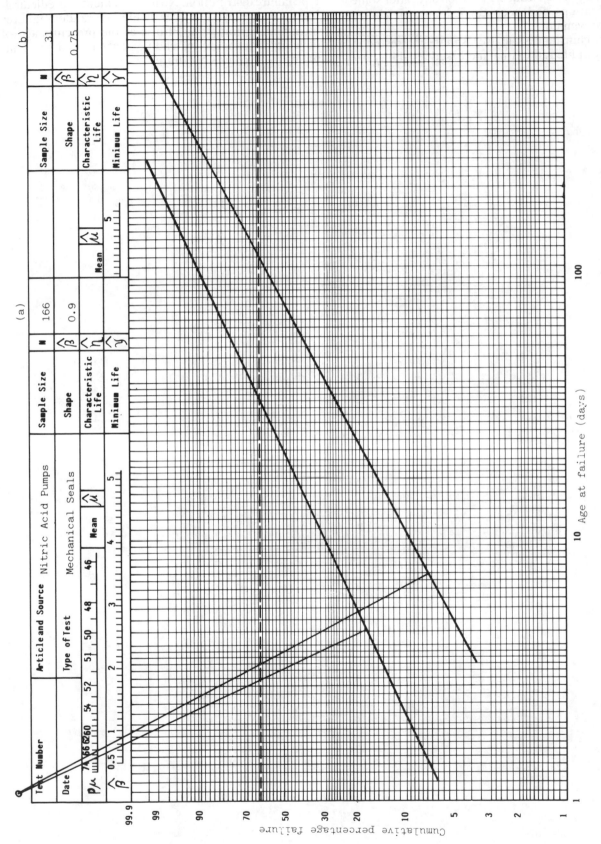

Fig. 1.23. Weibull plots for seal failures from nitric acid circulation pumps (a) before plant modifications, (b) after plant modifications

that, although there is little change in the Weibull index, the L_{10} life has gone up from 2.1 to 10.05 days after some plant modifications had been carried out, an almost five-fold improvement.

In many circumstances the user is more interested in predictability of life rather than increased reliability *per se*. Unfortunately this is incompatible with a random-type failure distribution. No improvement in predictability can be achieved until early life and random failures have been eliminated. Efforts at the present stage of development should instead be concentrated on eliminating these causes of failure.

Chapter 2

MATERIALS OF CONSTRUCTION
T. N. Cleaver

2.1 INTRODUCTION

This chapter describes the materials commonly used in the construction of mechanical seals and the factors influencing selection for different applications.

A mechanical seal can be split into four main groups of components:

(1) Pump related components – seal plate, etc.
(2) Face-loading elements – spring(s), bellows.
(3) Secondary sealing elements – 'O' rings, gaskets.
(4) Seal faces.

While individual seal designs vary, in almost all cases these components can be readily distinguished (see Fig. 2.1).

Selection of the materials used in mechanical seals is based on experience, backed up where necessary by laboratory testing. In normal circumstances material selection is the responsibility of the seal manufacturer and it is thus essential that he is provided with full details of the fluid coming into contact with the seal. These should include minor chemical constituents and off design conditions that can occur in service. For instance, pumps operating in the food or pharmaceutical industries may be subject periodically to special sterilisation procedures which can involve flushing with different chemical fluids from those normally handled.

Care should also be taken when the secondary sealing elements can be in contact with two different fluids. This happens in tandem or double seal configurations where

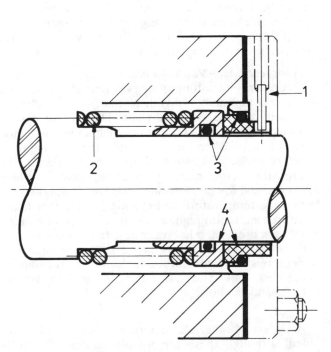

Figure. 2.1. General arrangement of mechanical seal showing the main groups of components.

some seals will be in contact with the pumped product as well as the barrier fluid.

The possibility of reactions occurring within the seal cavity with the formation of deposits on the seal, particularly on the atmospheric side, is an important feature that influences seal selection and further emphasises the need for complete details of the chemical composition to be provided, if satisfactory material selection is to be achieved.

2.2 GENERAL CRITERIA

2.2.1 All component groups

All seal materials that come in contact with the pumped fluid must:

(1) Withstand the prevailing temperature without significant loss of strength or degradation;
(2) Be mechanically strong enough to withstand prevailing pressures and rotational speeds;
(3) Be capable of withstanding any corrosive attack by fluids with which they come into contact.

2.2.2 Pump related components

Stainless steel 18/11/3 (AISI 316) is the normal material used for the pump related components. On corrosive duties the more resistant materials such as Duplex stainless steels, Carpenter 20-Cb3, Monel and Hastelloys are available to meet specific requirements.

In special cases the material to be used may have to comply with a user specification; for example, NACE MR-01-75 for petroleum service where sulphides are present.

2.2.3 Face loading elements

Stainless steel 18/11/3 is the normal material used, but again more corrosion resistant materials are available where necessary.

While stainless steel is suitable for low temperatures, at temperatures greater than about 250°C, an alternative material must be used because of relaxation and loss of face loading. Thus Inconel X750 is often used as a high temperature spring material. In the case of bellows loading elements a variety of materials is available and the seal manufacturer should be consulted.

2.2.4 Secondary sealing elements

The functions of the secondary sealing elements are:

(1) To seal 'static' leakage paths.
(2) To provide a degree of resilience to both rotating and stationary elements.

It is necessary for effective sealing that the seal faces are kept in contact by the loading element(s). As a consequence, one of the secondary seals must be capable of

Table 2.1.   Typical Properties of Mechanical Seal Secondary Sealing Materials

Type	Temperature (°C) Min	Max	Example of use
Elastomers:			
Nitrile rubbers	−30\−40	100\120	General purposes
Ethylene Propylene rubber	−50	140	Hot water, acids, alkalis, alcohols, ketones.
Fluorocarbon rubber	−15	180\200	Hot hydrocarbons
Less Common:			
Low nitrile rubber	−55	100	Some LPG applications
Fluorosilicone rubber	−55	200	Special chemical applications
Silicone rubber	−55	200	Special chemical applications
Perfluoroelastomer	−30	260	Special chemical applications
Non Elastomers:			
PTFE	−100	250	Low and high temperatures
PTFE/Composite	−100	300	High and low temperatures
CAF	—	500	High temperature
Compressed carbon	—	500+	High temperature

Note: These temperature limits are illustrative; the actual value may be affected by the sealed fluid properties.

axial movement, yet still retain its sealing function. (In the case of a bellows seal the axial movement is provided by the bellows, obviating the need for a dynamic secondary seal). Secondary sealing elements can occur in a variety of forms, such as flat gaskets, 'O' rings, V rings, wedges and U cups, the last three being mainly confined to the sliding element. 'O' rings can be used in both sliding and static applications while flat gaskets are restricted to static applications.

As one of the functions of the secondary seal is to provide a degree of resilience, elastomeric materials are the first choice up to the limits of their thermal and chemical resistance. A guide is given in Table 2.1 to the temperature limits and fields of application of the materials in common use. More detailed information is given in the following sections to draw attention to particular properties that influence selection.

Elastomers

Nitrile – Buna N (NBR)

Nitrile rubbers are polymers of acrylonitrile and butadiene with acrylonitrile content ranging from around 18 per cent to 50 per cent giving rise to a range of grades with different properties. Chemical resistance increases with increasing acrylonitrile content, but at the expense of low temperature properties. Medium to high nitrile grades are generally used; low nitrile grades are more suitable for low temperature applications.

Nitrile rubbers have good physical properties in terms of resilience, strength and abrasion resistance. They are generally resistant to mineral oils and greases, water, and many other chemicals. They have poor resistance to ozone and, as with most rubbers, care should be exercised in storage. They are not suitable for halogenated hydrocarbons, ketones, strong acids and certain hydraulic fluids.

Ethylene propylene (EPM or EPDM)

Ethylene propylene rubber is a co-polymer of ethylene and propylene, sometimes with a third monomer (terpolymer grades). These particular elastomers were originally introduced because of their excellent resistance to phosphate ester type hydraulic fluids. They have good mechanical properties with a wider range of temperature resistance than nitrile rubbers, and are used extensively on hot water duties. They are, however, severely attacked by petroleum based fluids – a fact which must be remembered when considering the use of grease as an aid to fitting mechanical seals.

EPDM is a similar material that provides superior properties in certain applications.

Fluorocarbons – Viton, Fluorel (FPM)

Fluorocarbon elastomers are polymers of two or more fluorine compounds, vinylidene fluoride, hexafluoropropylene and tetrafluoroethylene. Their major characteristics are their high temperature capability and wide range of chemical resistance. Within the working temperature range of elastomers already considered, the fluorocarbon elastomers are inferior, have a poorer resilience and lower abrasion resistance. It is, however, their wide temperature and chemical range that make them so useful, although, unless specially compounded grades are used, their hot water resistance is poor.

Fluorocarbon elastomers have excellent resistance to mineral oils and hydrocarbon fuels and are resistant to many chemicals. They are not resistant, however, to ketones and alcohols.

Others

A variety of other elastomeric materials can be provided, although these are usually confined to fairly specialised applications. A relatively new development is the perfluoroelastomers, typified by Kalrez, which are claimed to combine the chemical resistance of PTFE with the resilience of elastomers. At present these are expensive and, because of the highly specialised production techniques required in their manufacture, likely to remain so for some time. Nevertheless, they still can be extremely useful as "problem solvers".

'O' rings can be coloured to identify different elastomers. On certain applications, however, the physical properties of the elastomer may be badly affected with a loss of chemical resistance. While the process of colouring is being developed, it is worth checking with the supplier when using coloured 'O' rings on severe chemical duties or when operating near the temperature limits of the elastomer.

Non-elastomeric materials

PTFE – FEP

PTFE (polytetrafluoroethylene) is a plastic with outstanding chemical resistance. It is, however, far less resilient than elastomeric materials and has poor tolerance to thermal cycling. Nevertheless in many situations it is the only viable choice. The resilience can be improved by using a PTFE envelope energised by a coil spring.

Another development is the use of elastomer rings covered with FEP, fluorinated ethylene propylene, a close chemical relative to PTFE. FEP has similar chemical resistance to PTFE; it is a thermoplastic and has a lower high temperature limit.

High temperature materials

On high temperature applications no elastomeric material is suitable and an alternative has to be found. Compressed asbestos fibre, (CAF), has been used for some years, but now is rapidly disappearing because of health problems associated with the processing of asbestos. Various substitutes exist, although many of these have lower high temperature limits then CAF. Moulded exfoliated graphite foil is now being increasingly used on high temperature applications.

A problem with these materials is that they have very little resilience and are usually used as flat gaskets; special designs have to be used when they are applied in dynamic applications.

2.3 FACTORS AFFECTING SEAL FACE MATERIAL SELECTION

As we have seen in Chapter 1, Section 1.2.3, most seals operate in the mixed lubrication regime where the load between the faces is carried partly on a fluid film and partly by solid contact The seal face material has an important influence on film formation and, moreover, has to be capable of operating with some degree of solid contact without excessive wear or deterioration.

In addition to adequate chemical and thermal resistance, the following operational factors have to be considered in seal face material selection.

2.3.1 Pressure

The sealed pressure influences face material selection in several ways. High pressure tends to distort the seal faces and can, in the extreme, disrupt the film between them. This is a particular problem when using materials with a low stiffness. Pressure also affects the wear life of a seal as well as the heat generated at the seal faces.

2.3.2 Speed

Since one of the seal faces revolves with the shaft, it is important that materials selected are strong enough to withstand the centrifugal forces set up.

2.3.3 Lubrication and surface contact

The combined effect of pressure and speed influences material selection in a different way.

Although in some cases mechanical seals operate with the faces completely separated by a fluid film, in the majority of cases some contact takes place and boundary lubrication conditions exist. Even with seals in which the faces are normally separated, face contact occurs at starting and stopping, and may occur during transient pressure surges. Face combinations differ in their ability to cope with such conditions. A seal face pair that includes a self lubricating material (e.g., carbon, PTFE) gives some protection against surface damage in the event of contact, but such materials have poor resistance to wear by abrasion in the presence of solids. In this case a pair of hard, abrasion resistant materials may give a better life, although it imposes the need to maintain adequate lubrication conditions during operation.

The *PV* factor (see Chapter 1, section 1.3.1) is widely used as an indication of the severity of the duty and the degree of face contact. The ability of face material combinations to tolerate such contact may be expressed by a *PV* limit and in some cases a *PV* limit is used to set duty limits for a satisfactory wear life. However see cautionary note below.

Pressure and speed combined also influence the amount of heat generated by friction at the seal face. If this heat is not effectively dissipated to the surrounding product, there is a danger of disruption of the film by vaporization. One of the factors affecting the heat dissipation is the thermal conductivity of the seal rings themselves, high conductivity materials allowing better heat transfer to the surrounding fluid. These considerations can be particularly important when sealing liquids close to their boiling/bubble point.

Note: PV values should be treated with some caution. Different seal manufacturers may use *PV* values in slightly different ways. Also different methods of calculation may be used. Furthermore, allowable *PV* is a function of seal design as well as of the face materials.

2.3.4 Product physical properties

Fluid viscosity affects the heat dissipation from the faces to the surrounding fluid. The bubble point (mixtures of fluids) or boiling point (single component fluids) dictates the product temperature margin (ΔT) available for effective heat dissipation. As well as affecting heat transfer and heat generation, highly viscous liquids can impose high mechanical stresses on components, including the seal faces. Various other physical properties affect heat transfer from the seal faces.

2.4 THE SELECTION OF SEAL FACE MATERIALS

Seal faces must be selected as a pair. The following sections describe the materials in common use, their significant properties as seal faces and the counterface materials against which they are used. Typical properties are summarised in Table 2.2.

2.4.1 Carbon–graphite

The term 'carbon–graphite', or more colloquially 'carbon', is used for a range of carbon composites that are generally the first choice for one of the seal faces.

Advantages

(1) Good lubricating qualities under dry or boundary lubricating conditions.
(2) An ability to bed in quickly and take up any slight imperfections in face geometry.
(3) Good all round chemical resistance.
(4) Wide temperature resistance ranging from cryogenic temperatures to 250°C; this upper limit can be extended to 350°C by using certain metallised grades or to 450°C plus with electrographite grades.
(5) Reasonably strong in compression.
(6) Relatively low in cost and readily available.

Disadvantages

(1) Low tolerance to the presence of abrasives or crystallising liquids.

(2) Some chemicals attack either carbon itself or the impregnant; with strong oxidising agents (e.g., nitrates, chlorates) there is a possibility of hazardous chemical reactions.
(3) Not as stiff as metals and ceramics and so tends to distort at higher pressures.
(4) Some applications will not tolerate the risk of carbon dust entering the process; this is generally a hygiene requirement rather than a potential hazard.
(5) While strong in compression carbon grades are weak under tensile stress.
(6) Relatively easily damaged and care is required in handling.
(7) Low thermal conductivity.

Grades in use

The term carbon–graphite covers a wide range of different products. Carbon exists as two allotropic forms: diamond and graphite. The term carbon usually refers to such products as coke, charcoal or lamp black, which are sometimes described as amorphous carbon, though they are generally considered to be crystalline forms of graphite. It is the graphite with its hexagonal layered molecular structure that gives carbon–graphite its self-lubricating properties, while the carbon imparts strength.

The precise manufacture of carbon–graphite grades is a commercial secret, but generally the process consists of producing a base grade by mixing carbon in some form, i.e., lamp black or coke, with natural or artificial graphite and pitch or resin to act as a binder and hold the mixture together. After mixing and forming into a suitable shape, the compact is baked at about 1000°C. At this temperature the binder is converted into coke and holds the mass together. As a consequence of the baking operation the 'base grade' carbon–graphite is porous and requires impregnation to provide an impermeable

Table 2.2.   Typical physical and mechanical properties of commonly used face materials

	Carbon-graphite resin impregnated	Carbon-graphite antimony filled	PTFE 25% glass	Stellite 1	Ni-Resist	Aluminium oxide 99.5%	Tungsten carbide Co binder	Tungsten carbide Ni binder	Silicon carbide reaction bonded	Silicon carbide sintered
Density (kg/m³)	1800	2500	2250	8690	7300	3870	14700	14700	3100	3100
Youngs modulus (GN/m²)	23	33	—	248	96	365	630	600	413	390
Bending strength (MN/m²)	65	90	—	—	—	320	1750	1700	500	450
Tensile strength (MN/m²)	41	48	12–20	618 (UTS)	200	—	—	—	—	—
Thermal conductivity (W/mK)	9	20	0.4	15	40	30	80	70	200	70
Hardness	90–100 Shore A	85–95 Shore A	70–75 Shore D	600 HV	150 HV	1800 HV	1500–1600 HV	1300–1500 HV	2500–3500 HV	2500 HV
Thermal expansion coefficient (per °C × 10⁻⁶)	3.0	3.5	44–92	11.3	19.0	6.9	5.1	4.8	4.3	4.8

material and to give good running properties. For high temperature applications the baked compact is further heated to over 2500°C to produce electrographite grades.

The choice of the base grade, together with the type of impregnation used, depends very much on the application, with certain specialised applications demanding specialised carbon grades.

Three broad categories are used:

(1) Resin impregnated grades.
(2) Metal impregnated grades.
(3) Electrographite grades.

Resin impregnated grades have a wide range of chemical resistance; while electrographites may be more resistant, the resin impregnated grades have better wear properties.

Metallised grades can be impregnated with a variety of metals, such as antimony and babbitt. The chemical resistance of such grades is less than the resin grades, restricting their use in acid media. They are, however, stronger than plain grades and, against certain counterfaces, give better running performance (ability to cope with boundary lubrication conditions). It is also true, however, that against certain counterfaces, metallised grades give a poorer performance, often due to an electrolytic action causing the metal impregnant to 'smear', resulting in high wear.

Electrographitic grades are used when a high temperature capability (greater than 250°C) and better chemical resistance are required. Electrographite grades are not as strong as either plain or metallised carbons and, being relatively soft, suffer badly in abrasive conditions.

Counterfaces

The ability of carbon to run against a wide range of counterfaces accounts for its extensive use in mechanical seals.

2.4.2 PTFE

The self lubricating properties of PTFE would appear to make it a good candidate for a seal face material; however, because of its low strength and tendency to creep its use is restricted.

Advantages

(1) Good lubricating qualities.
(2) Almost total chemical inertness, though this can be reduced by having to add a filler, e.g., glass fibre, to improve its mechanical properties.

Disadvantages

(1) PTFE has a low strength and deforms easily under load. This can be improved by compounding with chopped glass fibre, but even so it has relatively poor properties compared with carbon grades. Its use is thus confined to relatively low duties.
(2) Despite its self lubricating properties PTFE does not perform well under boundary lubrication con-

ditions; high heat generation can cause severe deformation and lead to rapid failure.
(3) Limited use on abrasive applications.
(4) Low thermal conductivity.
(5) Relatively expensive.
(6) Limited use on pure water.

Counterfaces

PTFE is usually run against high purity aluminium oxide (99.5% Al_2O_3), giving a combination that is highly resistant to a wide range of chemicals, including those that attack carbon, but limited in application because of the low thermal conductivity of both faces.

2.4.3 Ni-Resist

Ni-Resist is a general term for a family of alloy cast irons containing nickel and other elements that give superior chemical resistance compared to cast iron.

Advantages

(1) Relatively cheap.
(2) Easily machined.
(3) Moderate corrosion resistance.
(4) High thermal conductivity.
(5) Good frictional characteristics, particularly in transient dry running conditions.

Disadvantages

(1) While better than cast irons, Ni-Resist irons still have relatively poor impact strength.
(2) While a relatively cheap face material, it generally has a shorter wear life than more expensive materials and is confined to relatively low duties.

Counterfaces

Ni-Resist cast iron is generally run against resin impregnated carbon–graphite.

2.4.4 Stellite

The Stellites are a group of alloys of cobalt, chromium, and tungsten with high hardness and excellent chemical resistance. They can be used as cast components or as weld overlays. While relatively cheap face materials they are now tending to be superseded in mechanical seals by the more expensive hard carbides.

Advantages

(1) A relatively cheap, hard material.
(2) Good chemical resistance, although when deposited on a base metal the latter determines the overall resistance.
(3) Good temperature range, maintaining hardness even at elevated temperatures.
(4) As a deposited material can be useful for large size seals.

Disadvantages

(1) Poor running properties on water or aqueous solutions when compared with other materials, particu-

larly when run against resin impregnated carbon–graphite.

(2) While initial cost may be less than other face materials, Stellite in general has a shorter wear life than more expensive materials.
(3) Low thermal conductivity.
(4) Very poor in dry running conditions.

Grades in use

Both Stellite 1 and Stellite 6 are used for seal faces with Stellite 1 the most popular.

Counterfaces

Stellite is normally run against metallised carbon–graphite; this gives a better performance than resin impregnated carbon–graphite, particularly on water and aqueous solutions.

2.4.5 Aluminium oxide

Alumina (aluminium oxide) ceramics were the first hard non-metallic materials to be used for mechanical seal faces.

Advantages

(1) A cheap hard material, very cost effective in large volume. Excellent wear resistance.
(2) Very good chemical resistance, depending on the grade used.
(3) Very good running properties on water and aqueous solutions using a carbon counterface. Can withstand mildly abrasive solutions, such as silty sea water.

Disadvantages

(1) Poor thermal conductivity, giving poor heat dissipation in critical applications.
(2) Poor thermal shock resistance; this can cause problems during transient conditions. Very poor dry running characteristics.
(3) A brittle material that is susceptible to mechanical damage.

Grades in use

Aluminium oxide ceramic is available in several different grades defined by the percentage of aluminium oxide. The impurities present are 'glass' or silica type impurities which can be attacked by certain chemicals, e.g., hydrogen fluoride, hydrofluoric acid, strong alkalis. The preferred grade for maximum chemical resistance is 99.5% Al_2O_3.

Counterfaces

Alumina is generally run against resin impregnated carbon–graphite or filled PTFE, the latter being used for highly corrosive chemical conditions.

2.4.6 Cemented carbides (hard metals – tungsten carbide)

Cemented tungsten carbide consists of hard carbide particles bonded together by a ductile metal. Traditionally it has been used on more severe duties (in terms of *PV*). While more expensive than the materials already discussed, it is now being increasingly used on less severe duties because of the improvement in seal life.

Advantages

(1) Good wear properties on more severe duties.
(2) High thermal conductivity.
(3) High elastic modulus, hence less prone to pressure distortion then metallic face materials.
(4) Better resistance to mechanical shock compared with other hard non-metallic materials.

Disadvantages

(1) Limited chemical resistance, particularly on acid duties.
(2) Very high density material; this can be critical on high speed rotating applications.
(3) Limited ability to cope with dry running conditions or boundary lubrication conditions when run against itself.
(4) High raw material cost.

Grades in use

Cobalt and nickel bonded tungsten carbides are most commonly used for mechanical seal faces.

The binder phase provides toughness and tensile strength and it is generally the binder that dictates the chemical resistance. Tungsten carbide grades are poor in acid media; cobalt grades being restricted to pH values above 7. Nickel binder grades have improved chemical resistance, particularly on water and aqueous solutions, but are still restricted to pH values above 6. Special binder grades that are resistant to pH values as low as 2 are available, but these are relatively expensive. The greater chemical resistance of tungsten carbides can give them an advantage over certain silicon carbide grades in alkaline conditions.

Counterfaces

Tungsten carbide is normally run against resin impregnated carbon–graphite or, where extra mechanical strength is required, against metallised carbon grades. When run against carbon, tungsten carbide is a good counterface and less susceptible to thermal shock than aluminium oxide. The combination is good with regard to transient conditions and under conditions of boundary lubrication.

In abrasive media, tungsten carbide can be run against itself or against silicon carbide. In this case tolerance to dry running and boundary lubricating conditions is poor.

2.4.7 Silicon carbide (SiC)

Silicon carbide is now becoming widely used, not only in

high duty applications, but even on lower duties as the benefits outweigh the higher initial cost.

Advantages

(1) Good wear resistant and frictional properties on severe duties.
(2) High thermal conductivity, comparable with and, in some cases, better than tungsten carbide.
(3) Good thermal shock resistance.
(4) High elastic modulus.
(5) Good chemical inertness.
(6) Lower density than tungsten carbide.
(7) Lower in cost than tungsten carbide.
(8) Raw material plentiful.

Disadvantages

(1) Lower toughness, depending on the grade, than tungsten carbide; can be easily damaged mechanically.
(2) Low strength in tension which, to some extent, offsets the density advantage over tungsten carbide.
(3) Certain grades are attacked by strong alkalis.
(4) Care has to be used in selecting grades to run together. The use of the wrong grades can result in high heat generation at the seal faces that could cause vaporisation of the interface film and face damage.

Grades in use

Care should be taken in selecting silicon carbide; the differences in properties between the grades from different suppliers are greater than with respective tungsten carbide grades. There are three families of silicon carbide.

Sintered alpha

These grades contain no free silicon. They have the best chemical resistance, but lowest fracture toughness. The friction characteristics are poorer than reaction bonded grades, but superior to tungsten carbide.

Reaction bonded

These grades contain free silicon and have the best friction characteristics of all superhard seal face materials. Some acid or alkaline media can cause leaching of the free silicon, but they are generally more inert than tungsten carbide.

Converted

These grades consist of a carbon–graphite core with surface converted to form a silicon carbide 'case'. They can be cost effective in high volume as superior materials to aluminium oxide, and can be run as a face pair for abrasive applications.

The mechanical properties are not as good as solid SiC grades because of the carbon core. They are used mainly on lower duties.

Counterfaces

Silicon carbide is usually run against resin impregnated carbon–graphite, although metallised carbons can be used for high performance, including hot duties. Silicon carbide against carbon is a frequently used combination for long life in a wide variety of conditions, because of its excellent resistance to thermal shock, transient, and boundary conditions.

In abrasive applications silicon carbide is normally run against tungsten carbide, giving the most effective combination for wear resistance and friction. Silicon carbide can be run against itself for extremely abrasive conditions, but frictional characteristics are not as good as silicon carbide versus tungsten carbide. When silicon carbide is run against itself the best results are obtained by using different types, e.g., sintered alpha against reaction bonded.

The combination of silicon carbide against tungsten carbide has been used successfully on high duty applications where carbon causes problems because of high distortion and wear. In such cases, however, the face configuration has to be carefully designed.

Generally, however, as with all hardface combinations, boundary conditions can result in surface thermal shock and care has to be taken to avoid even transient dry running.

2.4.8 Other face materials

Since the seal faces are vital to the functioning of the mechanical seal, new materials are constantly being evaluated.

Where cost is of prime importance and duties are very low, materials such as stainless steel and lead bronze can also be used. Chromium oxide is used as a deposited coating on stainless steel. There is, however, a danger of the coating cracking or lifting off under dry running conditions.

2.5 ECONOMICS OF FACE MATERIAL SELECTION

The discussion so far has tended to relate face material selection to degree of severity of the application (usually defined as a *PV* limit). That is the duty dictates the face material according to severity. Materials like tungsten carbide and silicon carbide are more expensive in terms of initial cost, mainly because the basic raw materials are expensive and the processing and machining operations are more complicated. Nevertheless, it may well be worth using the more expensive materials when the advantages gained by fewer seal replacements could more than outweigh the initial cost.

To see why life could be extended three major factors can be considered:

(1) Wear rate.
(2) Face temperature.
(3) Surface flatness.

2.5.1 Wear rate

If other causes of failure can be avoided the life of mechanical seals is ultimately determined by wear of the seal faces. Hard face materials, such as tungsten and silicon carbides, are very wear resistant. This can be particularly advantageous where abrasive solids are present in the sealed fluid. In such circumstances carbon has a high wear rate and the benefit of increased life with the more expensive hard face materials may well offset their higher initial cost. Another benefit that may be gained through a reduced wear rate is the decrease in rate of production of wear debris (carbon dust) that could, ultimately, lead to seal 'hang-up' failure.

2.5.2 Lower face temperatures

As can be seen from Table 2.2, tungsten carbide and silicon carbide have higher thermal conductivities than other face materials in common use. The effect of this is to provide more effective dissipation of the heat generated at the seal faces and, hence, lead to lower face temperatures. This can have a beneficial effect in reducing the risk of vaporization of the interface film.

2.5.3 Surface flatness

When tungsten carbide or silicon carbide or even aluminium oxide are run against carbon, more uniform wear occurs than on some of the cheaper counterfaces. This means that there is less likely to be grooving wear that can lead to increasing leakage before the full wear life of the seal is achieved.

2.5.4 The cost of a mechanical seal

It is important to realise that the initial cost of a mechanical seal is not the full story. In fact the reduction in maintenance costs achieved by using a more expensive seal face material can result in significant life cycle savings as illustrated in the following example.

Example.

50 mm seal on water at 1.5 MPa.
Speed: 3000 rpm
Face combination: Stellite 1/carbon–graphite

Seal cost: £180
Replacement cost: £1000
Average life (MTBF):   7 months.

5-year maintenance cost: 8 seal replacements

8 × 180 + 8 × 1000 = £9440

Change to face combination reaction-bonded silicon carbide/carbon–graphite

Seal cost: £220
Average life: 14 months

Projected 5-year cost: 4 seal replacements

4 × 220 + 4 × 1000 = £4880

In this case the change of face materials is expected to reduce the maintenance cost by about 50 per cent over a 5-year period.

PART II

Mechanical seal selection

Chapter 3.

DATA REQUIREMENT FOR SEAL SELECTION

P. R. Rogers

3.1 INTRODUCTION

The seal manufacturer relies on knowledge of his technology, experience and, most important, the quality and extent of the information provided to him before selecting a mechanical seal. It is essential that the maximum amount of relevant information should be provided. This should not only cover the characteristics of the pumped fluid, but also the operating conditions and the availability of services which may be necessary to enable the seal to perform satisfactorily and operate reliably. Failure to provide this information carries with it the risk of an inappropriate selection resulting in premature seal failure and consequential hazards if the sealed liquid is toxic or flammable.

This chapter presents a data sheet covering the information required by the seal manufacturer in order that he can recommend a suitable seal arrangement and expands on the reasons why this information is necessary. The data sheet is the starting point of an enquiry; if needed if may later be supplemented by direct discussion between the purchaser and the seal manufacturer, but the information called for on the data sheet must be responded to in the first instance.

3.2 DATA SHEET

A recommended form of data sheet is given in Fig. 3.1. All the items shown can be important, although clearly some rank higher than others. The more the information on the details of the application that can be supplied, the better chance the seal manufacturer has of providing a seal that will perform effectively.

It is wholly inadequate to expect the seal maker to propose a satisfactory seal on the basis solely of a statement which gives the name of the liquid to be sealed, its pressure and temperature, together with the size of the shaft or shaft sleeve. Selection of the seal is only the first step. In addition, some control has to be exercised over the environment in which it will operate if it is to perform satisfactorily. The information necessary to make recommendations for a suitable seal arrangement has also to be provided on the data sheet.

The object of the data sheet is not, however, merely limited to the provision of fact; it can additionally be used to indicate the purchaser's preference based on his experience and operating constraints. With the data provided, the seal selector will place the technical options which are open within a framework of what is possible, practical and desirable from the point of view of the equipment, its operation and its cost.

Satisfactory selection and eventual operating experience starts from good data sheet information, and it is therefore important to provide as much as possible.

3.2.1 Data Sheet Structure

The data sheet, Fig. 3.1, is divided into broad sections covering different aspects of seal selection.

PART 1   *Commercial details and enquiry requirements*

PART 2   *Minimum seal selection data*
The source of this information may be the end user, pump manufacturer, or contractor. Seal size may be defined in different ways according to the manufacturer or a seal standard. For clarity, therefore, the shaft and sleeve size at the seal are requested.

PART 3   *Supplementary operating data*
This section covers the areas of concern which may affect the style or quality of the seal and materials offered. It may also significantly affect recommendations made on installation design and use. Where pumps may be used on alternative services, data for these should also be given.

PART 4   *Operational safety*
As much advice as possible is sought and the onus should not be placed upon the seal manufacturer to advise when a service may be hazardous. Chapter 4 covers this requirement in more detail.

PART 5   *National and international standards*
The seal makers have developed particular product lines and approaches to meet the requirements of the various standards.

PART 6   *Installation details*
This section brings in pump manufacturer and user application experience.

PART 7   *Seal design preference*
This again brings in pump manufacturer and user experience. The pump maker may have an established seal and installation standard or the user may already be using and possibly stocking a particular seal.

PART 8   *Secondary containment*
This section, which provides some of the standard leakage containment options, is associated with the hazardous nature of the duty specified in Part 4. If a double or tandem seal is specified in Part 6 any requirement stated in Part 8 would be installed on the atmospheric side of the outer seal. This topic is considered in Chapter 4.

PART 9   *Supporting services*
Knowing what services are available can have a considerable influence on seal selection and installation.

When requesting a mechanical seal selection, please complete the following questions as far as possible. Comprehensive data will assist the seal selector to provide equipment most appropriate to the stated need.

Part 1 - Purchaser Requirements:
Raised by .... Company .... Date ....
Pump service .... Plant item number .... Ref. Pump drgs. ....
Enquiry Ref. .... For Proposal/Purchase ....
Seal manufacturer .... Prop.req. by ....
Special certification ....Y/N   (ref part 12)   Seal Inst.Drg.Reqd....................Y/N

Part 2 - The minimum application details for provisional seal selection are:
Liquid ....
Temperature ....°C   Req. seal size ....mm/in Shaft/sleeve size under seal ....mm/in
Sealed press. ....Bar g   Speed ....(rpm) Rotation ...CW/CCW Rev.Rotation..Y/N @ speed max........rpm

Part 3 - The following information will improve the quality of the seal specification.
Pump suct.press.............(bar g) Pump disch.press......... (bar g)   Static press.max/min........ (bar g)
VP at PT. ....(bar abs) Bubble point ......(°C @ Sealed pressure) Vacuum ......Y/N.......... (bar abs)
Abrasives .... Y/N   Constituents........................... Concentration ....
Dissolved solids (constituents)...................................... S.G. ......   Corrosive/pH.. ...
Auto-ignition point ....(°C)   Max/Min. ambient temp. ......(°C) Chlorides.....ppm Hyd.Sulph.............ppm
Pour point ....(°C)   Viscosity norm./max. ....(cSt) Carb.Diox.....ppm Other ....ppm
Special Operating Comments/Requirements ....Dry Running....................Y/N

Part 4 - Is the liquid being handled hazardous in any way? .... Y/N.   Please provide full details.
Hazards (state)..
Toxicity Rating...............   Is leakage tolerable? ...... Y/N @ ......litres/hr.

Part 5 - Do you require the mechanical seal to comply with an established standard?..............................Y/N
BS .... DIN .... ISO .... ANSI ...... API .... NACE ....Other........................

Part 6 - Do you require a particular installation? .... Y/N
Single.........Y/N Double back to back ....Y/N Double face to face .....Y/N Tandem ....Y/N
Cartridge......Y/N Stationary mounted...........Y/N Can a clean flush be used? ....Y/N (Ref. Part 9)
Compatible sealant for double seal installations ....

Part 7 - Do you require a particular seal design/type? ... Y/N
Rubber bellows...Y/N   O-ring...Y/N   PTFE Wedge...Y/N   PTFE O-ring.......Y/N   Metal Bellows.......Y/N
Unbalanced/Balanced.   Single/Multi Spring.   Seal Materials (Specify)...............................

Part 8 - Do you need a containment/safety seal in addition to the installation stated in Part 6....................Y/N
Non Spark bush .. Y/N   Lip seal.. Y/N   Labyrinth bush.. Y/N   Mechanical.. Y/N   Floating labyrinth .... Y/N
Standstill ..... Y/N   Other.................. Max. Press ..... (bar g) Max. temp. ..... (°C)

Part 9 - What on-site services are available?   Please give details.
Water ... Y/N @ Pressure......(bar g) & Temperature.....(°C)
Steam ... Y/N @ Pressure......(bar g) & Temperature.....(°C)
Flush ... Y/N   Liquid .... @ Pressure .... (bar g)
                                        & Temperature ....(°C)

Part 10 - Do you require the seal supplier to provide any associated items of equipment? ... Y/N
Sealant System ... Y/N (att. scope, spec. & stds) ref.Pt.6. Cooler ... Y/N Type ....(Standard ....)
Cyclone Separator... Y/N   Filter ... Y/N   Preferred Type ....
Flow Controller ... Y/N Long/Short Leakage detector....Y/N Other...................................

Part 11 - Details of the sealed equipment help to determine the seal installation.
Pump Maker ....,......... Pump Type .... Description ....
Horizontal/vertical .... Axial/Radial split   Seal mounted on shaft/sleeve .
Seals per pump .... Axial shaft movement .....(mm/ins) Driver............... Electric Motor/Steam Turbine/Engine.
Wetted parts materials..

Part 12 - Do you require any special material certification or seal material/performance test? .... Y/N
Specify Certification ....Seal Tests (Std/Spl) ....

Fig. 3.1. Mechanical seal specification – Duty Specification

The information requested will enable seal installation drawings to be produced for order. If a full cartridge is required please give details of sleeve ends and locking arrangements together with pump assembly/maintenance instructions, wherever possible please provide detail drawings of the seal chamber/existing chamber connections/shaft and sleeve. These will assist first time installation drawing accurancy and quick response to your enquiry.

The following dimensions relate to the sketch below. The upper half represents a typical overhung impeller centrigugal pump. The lower half is for a between bearing pump. For other types of pump please produce a working sketch.

All dimensions in mm/ins

D1	Shaft or sleeve *................	Lz	Nearest obstruction (axial) ....
D	Shaft ∅ under sleeve *...........	Dz	Nearest obstruction (radial).........................
D4	Seal chamber bore *..............	No of seal plate bolts .... Size ....
D4+	Max. seal chamber bore *........	PCD .... °Offset from vertical C/L ....
D53	Spigot *........................	M	Max shaft movement (assembly) ....
L42	Seal chamber depth		(operation)........................
Lx	Sleeve extension	Lt	Chamber face to shaft step..........................
Ly	Stud protrusion	Lu	Chamber face to sleeve end ....
Lv	Max............ Min	L30	Chamber face to connection C/L......................
Pump ref. drawings......................		Connection size ....
....	C/L	Shaft distance (twin screw pumps) ....

Other dimensions ....

* Tolerances required

\+ For stationary mounted installations provide max. possible chamber bore for the working pressure.

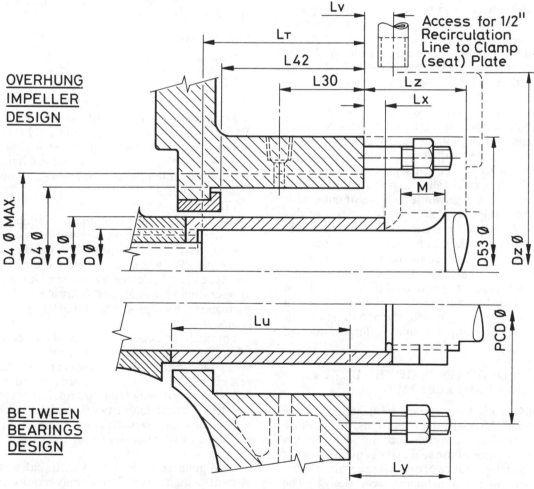

Fig. 3.2.   **Seal installation dimensions.**

PART 10 *Ancillary equipment*
This again is a pump maker's or user's preference section.

PART 11 *Sealed machine information*
This is important background so that a seal with the right design characteristics is selected for the appropriate installation.

PART 12 *Test and inspection requirements*
A clear understanding of requirements may affect the particular style of seal selected, its constituent materials and cost. Seal testing and verification of seal selection is covered in Chapter 8.

3.2.2 Installation Dimensions

For installation drawings and conversions from one form of seal to another, a number of essential dimensions are required for accurate production drawings. The nomenclature used where possible is that also used by a number of European seal standards. The dimensions and basic sketches for overhung and between bearing pumps are provided in Fig. 3.2.

Having relevant dimensions available while selecting a seal makes it possible to consider alternative seal selection and installation options. Besides the possibility of cartridge seals, which can reduce initial fitting problems, there is also an increasing interest in stationary seals, viz, designs in which the floating seal member is mounted on the seal plate. This can reduce the problems that arise from misalignment, imbalance and externally imposed vibrations, and can increase overall seal life. This particularly applies to metal bellows seals.

One of the considerations in seal safety and reliability is that seals should retain integrity on bearing collapse. Stationary mounted floating seals can provide this to some degree, because shaft-to-seal component clearances can be made to suit the particular application.

However, this type of installation may require a larger seal than one that is conventionally shaft mounted. Some standard pumps may not be able to accept this without an increase in the seal chamber bore. The size of the seal then becomes a key factor in seal selection.

Another factor, often overlooked, is shaft movement during assembly or operation. Unlike gland packing, mechanical seals have restricted axial movement. If the shaft float exceeds that built into a particular seal design then another seal must be chosen, the first design modified to suit, or the pump assembly tolerances/design amended.

3.3 THE SIGNIFICANCE OF THE DATA TO SEAL SELECTION

The data sheet calls for detailed information on the sealed liquid and the operating conditions. This section considers the way these can influence the seal selection process and the choice of the seal arrangement, in order to emphasize the importance of providing accurate information in obtaining a satisfactory seal design. The influencing factors are discussed separately for the sake

of clarity, though in the final analysis it is the combination and interaction of the different factors which is crucial.

3.3.1 Sealed liquid

Chemical properties

Detailed information on the chemical properties of the sealed liquid is essential for correct seal selection, both for the effects the liquid may have on the seal parts with which it comes into contact and also for any secondary effects arising from leakage. The seal parts have to be resistant to chemical attack, but possible deleterious physical effects on any elastomeric secondary seals (e.g., swelling) have also to be taken into consideration.

Minor or trace constituents can be significant (e.g., chlorides can cause corrosion of 316 stainless steel, a material commonly used for the structural parts of the seal), and it is important that full details of these are included in the specification of the sealed liquid.

Although the leakage from mechanical seals is normally small (see Chapter 1, section 1.4.1), any formation of deposits on the outside of the seal caused by reaction of the leakage with the atmosphere (e.g., coking with hydrocarbons), or crystallization of dissolved solids through evaporation can influence the choice of the seal or seal arrangement.

Physical properties

Vapour pressure
In seal selection, vapour pressure is considered in much the same way that NPSH is used in pump selection. It is not a performance measurement, but it is an essential consideration in the correct operation of the seal. The provision of an adequate margin between the temperature of the product in the seal chamber and its boiling or bubble point is discussed in Chapter 1, section 1.3.2, Product temperature margin (ΔT limit).

Viscosity
Although centrifugal pumps are limited by viscosity considerations, mechanical seals are also used on positive displacement pumps which handle much higher viscosities. The whole subject is a tribological study, but, briefly, the seal selector has to consider starting as well as operating viscosities and be aware that some liquids, particularly food products, have very peculiar viscosity characteristics.

These shear dependent viscosity effects have to be taken into account as they may influence the performance of the seal, both with respect to heat transfer in the seal chamber and the formation of the interface film.

The sealing of centrifugal pumps has been the basis for the design of the majority of standard mechanical seals and in consequence such seals are normally satisfactory for liquids with viscosities that can be handled by such pumps.

As a general guide, however, liquids with operating viscosities higher than 750 cSt may require modifications to seal design or installation. Some seal designs are more capable than others at handling high viscosities and the newer designs of metal bellows seals are very good in this respect.

High viscosity can, if not designed for, result in broken seals. This normally occurs on start up because the pumps have not been warmed through or abnormally cold conditions apply. Although not strictly a viscosity effect, atmospheric side icing can have the same effect as a highly viscous liquid on start up.

Viscosity, therefore:

(1) determines whether standard or slightly modified seal components can be used – this may mean as little as narrowing the faces;
(2) has an influence on the seal design proposed;
(3) has a direct influence on the style of seal counter-face (seat) by requiring a positive anti-rotation feature, such as a pinned or a clamped seat;
(4) may suggest the need for discussion with the pump maker about, for instance, the need for pre-heating built into a seal installation;
(5) will directly affect the starting torque and power consumption of the seal.

Note. Seal manufacturers have published power and torque figures which are based upon research data, but their application to fluids with non-Newtonian characteristics must be handled with care.

Fluid specific gravity

This does not directly affect seal performance. It does however provide an indicator to other parameters which do.

Low specific gravity of hydrocarbons suggests high volatility and can be used as a check on vapour pressure data.

Conversely, a high aqueous solution specific gravity suggests a high level of dissolved solids, details of which should be looked for in the appropriate part of the data sheet.

Seal manufacturers invariably consider pressure, but pump performance data is often given in terms of head. Specific gravity is needed for conversion between the two.

Specific gravity is therefore used to:

(1) highlight incorrect data;
(2) convert pressure and head.

Free solids in the pumped fluid

Mechanical seals are surprisingly tolerant of undissolved solids. These may occur in saturated solutions or as insoluble debris. Slurries may contain soft or hard particles. Their shape may affect how abrasive they are. Sludges and stocks may contain fibrous materials and both the seals and their installations may require features to stop the seal chamber and recirculation line becoming plugged with fibres to the extent that seal flexibility is impaired.

Small bore piping typically used for recirculation may plug and reduce the flow. Flow restriction devices in recirculation lines are particularly prone to blocking. In either case the reduction of cooling available to the seal can lead to seal failure.

Any cooler to be fitted in a recirculation line may need careful design consideration.

Large solids tend to cause impingement damage on the liquid side of the seal, so the method and position of seal recirculation in relation to the seal needs consideration. Because the seal faces form a very effective edge filter, it is only small particles which can enter the fluid film early in the life of a seal to cause wear. Sub-micron particles tend not to cause too much of a problem. It is those about the same size as the film, 0.5–2 μm, which cause most damage.

Another consideration is when there is a need to filter the liquid being recirculated to the seal. A cyclone separator is preferred because it is less likely to clog, but it is essential to have a significant difference in specific gravity between the particles and the carrier liquid for it to be effective.

Process systems can be very dirty during and immediately after commissioning, leading to the so called settling down period. 'Once through' systems tend to clean up much quicker than 'closed loop' systems in which pipe scale, sand, and other debris have to find points of low turbulence to settle out.

A knowledge of solids content, therefore, leads to:

(1) seal type selection review;
(2) seal face material review;
(3) preferred seal installation in line with available services;
(4) possible changes to pump design in the seal area.

Dissolved solids

At first sight, solutions would appear to create no difficulties for the seal. However, if part of the leakage occurs by evaporation of the interface film on the atmospheric side of the seal, the solution becomes concentrated and may precipitate solids.

The effect of these solids depends on their physical properties. Hard crystals are potentially abrasive and may require the use of hard seal face materials to resist wear. Precipitated solids may bind the seal faces together during shutdowns. If deposited outside the seal they may inhibit the free flexibility necessary for the seal faces to remain in contact, causing hang-up. Deposits forming in drains can block them.

A knowledge of the dissolved solids can therefore be used to assess the need:

(1) to select highly abrasion resistant faces;
(2) to select seal features and materials which will minimize seal face operating temperatures (as it happens the hardest materials used by seal makers, tungsten and silicon carbide, also have high thermal conductivity, which supports this requirement);
(3) to select seal types most resistant to seal hang up;
(4) to suggest seal installation features which will minimize seal operating temperatures and restrict atmospheric seal side precipitation.

Hazards

Normal leakage of liquid or vapour from a seal may present a hazard to local staff or plant. Seal failures present a much more serious situation.

Hazards fall into three categories, arising from toxic, flammable/explosive, and corrosive materials.

Hazards should be defined by the user, but it is often the case that a seal selector's view of the potential hazard may well produce a seal specification and supporting requirements which calls for changes in both the pump and on-site services. This can introduce a cost element which may be more than the user feels can be justified economically.

A genuine understanding and good working association between seal maker, pump maker, and user is therefore needed on key equipment, safety standards, and the approach to be taken by the seal selector.

Leakage containment can be handled to a greater or lesser degree by devices such as quench bushes and lip seals of varying types, but their effectiveness depends upon many factors. Special designs of back-up mechanical seals are available from individual seal manufacturers. (See Chapter 4, section 4.6, for discussion on secondary containment.)

Double and tandem seals provide close to absolute leakage prevention and are, therefore, often specified for the most critical applications.

Hazard considerations are therefore used to define:

(1) the approach to be taken by the seal selector;
(2) the type of secondary containment device most appropriate;
(3) particular services which may be required for operational safety and remote leakage detection.

3.3.2 Operating conditions

Temperature

Temperature has a major influence on seal material selection, particularly the materials used for secondary seals and bellows, which have both high and low temperature limits. In addition to its effect on vapour pressure and the need to maintain a sufficient product temperature margin in the seal chamber, temperature can influence the design of the seal arrangement in a number of other ways.

Information on temperature is thus needed to:

(1) help determine what materials might be suitable;
(2) apply constraints to established seal operating performance data (the section on vapour pressure provides one example of this – see Chapter 1, section 1.3.2);
(3) indicate possible mechanical considerations affecting seal selection (for example, shaft expansion);
(4) suggest installation features (for example, low pressure saturated steam quench for hot hydrocarbons to prevent coking).

Pump suction, discharge, and sealed pressure

It is usual for both suction and discharge pressures to be defined for the pump selection process. Depending upon pump design, the sealed pressure, which is of most concern to the seal manufacturer, is usually somewhere in between the two and often has to be assumed by the seal selector. This may make the difference between offering balanced or unbalanced seals (see Chapter 1), which in turn may affect the pump being offered by requiring stepped or unstepped shafts or sleeves. The balanced seal (if not of the bellows type) in a single stage over-hung impeller pump requires a sleeve, whereas an unbalanced seal can be fitted to a straight shaft. This is just one practical example of how inadequate data can lead to incorrect seal selection and cost.

If operation at reduced flow is anticipated, sealed pressure may rise significantly and should be noted. This particularly applies to spared high pressure multi-stage pumps, when the second pump may be fully pressurized by the first before start up.

During manufacture, seals are not usually subjected to hydrostatic test pressure, but in practice there may be static pressure requirements outside the operating sealed pressure range. These must be stated. Alternative duties should also be indicated on the data sheet, or a separate data sheet completed.

Operating pressures are used:

(1) to select one or more seals with appropriate pressure ratings;
(2) in association with both seal size and speed, to ensure that the seal faces have adequate performance (see Chapter 1, section 1.3.1);
(3) to determine aspects of the installation, such as clamp plate design, type of recirculation employed, clean injection flush pressure, and desirable barrier liquid supply pressure to double back-to-back or face-to-face seals.

Shaft speed and direction of rotation

Mechanical seals can be called upon to operate from a few r/min to 50 000 r/min and sometimes higher on special applications. However, the majority of industrial pumps are induction motor direct driven at two or four pole motor speeds.

There are occasions when the drive will be taken through gear boxes or belt drives and, increasingly, electronic control is being used to produce variable speed drives.

Alternative types of driver, such as steam and gas turbines, power recovery turbines, and engines, are far less common than electric motors. For all types of variable speed driver, maximum speeds must be considered.

Shaft speed is considered:

(1) in conjunction with sealed pressure and shaft size, to establish that the seal faces have adequate performance;
(2) to ensure that the rotating elements of the seal remain functional (At high speeds the positions of the seal unit and its counterface may be reversed so that the unit components are not subjected to

centrifugal effects. This may also be size and seal type related and more likely to affect unshrouded single spring seals. One very different effect of this reversed or stationary mounted installation is that it is less susceptible to pump component misalignments and deflections if properly designed, and is particularly appropriate to metal bellows seal installations.);

(3) to establish trip or reverse running speeds.

Note. Direction of rotation is also a factor in seals driven through springs. For seals with this feature, it is often essential to select a seal for the right rotation. Such seals will be opposite handed for double ended pumps and should not be confused during maintenance.

3.3.3 External services available

On a number of occasions in this chapter reference has been made to cooling water, steam quench, and clean liquid injection. It is very useful to have information about these services when giving detail consideration to seal selection and also the pressures and temperatures at which these services are available. For instance, some form of cooling may be applied for hot seals. Start up heating may be required for very viscous liquids. Liquids which may deposit waxes or gums may require a steam quench to extend seal life.

3.3.4 Seal standards

Standards are an everyday feature of industrial life. There are company standards and specifications, national standards, some of which are well known internationally, and international standards. Some company standards, typically in the oil industry, use a national industrial standard as a basis.

Standards also differ in the way in which they achieve their objectives. API 610 – *Centrifugal pumps for general refinery services*, which is essentially a pump design feature standard, includes seal requirements and, in the latest edition, seal chamber dimensions. It is on this standard that many oil companies base their own individual standards.

Another US standard, ANSI-B73.1, is essentially a pump dimensional standard which includes some seal chamber dimensions.

DIN 24 960 is a detailed dimensional, rather than performance seal standard. It has been partly embodied in a highly modified form in ISO 2858 standard which, with the ISO standards on base plates and couplings, became BS 5257 – *Horizontal and suction centrifugal pumps*. Other European countries have taken the DIN standard as a basis of their own national standard.

A list of standards relevant to mechanical seals is given in Appendix 3. For most of these standards, seal manufacturers have developed particular seals which comply with the dimensional requirements.

These specifications are a powerful tool when used with proper understanding. However, when used indiscriminately they may not produce the desired effect and should not be substituted for detailed discussion of difficult sealing duties with the seal manufacturer to arrive at the best selling solution.

Seal standards may define:

(1) acceptable seal designs;
(2) seal cavity dimensions;
(3) acceptable materials;
(4) ancillary equipment.

3.3.5 Pumping equipment

Seals can be selected without any knowledge of the equipment but the interpretation of the information contained on the data sheet is governed to some degree by the seal selector's understanding of the equipment to which the seal will be fitted.

The general definition of the pump type or other machine involved enables the seal selector to interpret the data and advise how the seal installation can be designed and assembled.

Information on the equipment:

(1) provides an additional basis for interpretation of the seal requirements;
(2) governs to some degree the good design of seal installations;
(3) suggests possible operating requirements which may not be directly specified on the data sheet.

Chapter 4

SEAL SELECTION

J. M. Plumridge and R. F. Polwarth

4.1 INTRODUCTION

This chapter is split into two distinct sections. The first section deals with the primary seal, which is defined as the mechanical device that contains the pumped or compressed medium in industrial and process rotating equipment. This section describes the different designs of mechanical seals, their strengths and weaknesses, and the factors that influence seal type selection.

The second section considers the conceptual issues of safety, multi-seal systems, secondary containment, and environmental and leakage control systems. In safety terms, toxicity has been considered the dominant parameter, followed by explosivity, and then flammability.

It should be emphasised that seal applications engineering is complex. This chapter presents a very simplified guideline that can be easily interpreted and understood and covers the majority of sealing duties.

PART A SELECTION OF THE PRIMARY SEAL

4.2 SELECTION PROCEDURE

To the new user of mechanical seals there must seem a bewildering choice of products from the different seal manufacturers. It is the manufacturer who will have the experience and background of each of his own products and thus it is he who should have the primary responsibility for seal selections. Whilst there will be subtle differences in selection technique, the following procedure can be considered common to all manufacturers.

4.2.1 Input data

The seal manufacturer should receive a comprehensive Data Sheet (see Chapter 3, Fig. 3.1) from which selection can be made with confidence. There are three main areas to be considered.

Operational requirements (data sheet)

The dominant parameters here are seal housing envelope, pressure, speed, temperature, and medium. Section 4.3 considers each of these parameters in turn.

Performance expectation

The dominant parameters here are life and leakage expectation. In addition, the requirements for pre-delivery proof testing and its acceptance criteria must be specified.

Special requirements

Requirements not covered by the above. For instance, approval authority, timescale, packaging, etc.

4.2.2 Consideration of duty

Envelope

It is necessary to have full dimensional data of the seal housing, any limitation on positioning within the housing, distance to the nearest external axial obstruction, and location of external equipment in the way of the seal housing and seal plate.

From this information the seal manufacturer can establish the seal size, check the ability to fit the seal arrangement without housing modification, and judge whether there is adequate clearance space to prevent fouling and/or to provide adequate flow paths.

Pressure

It is important to have a reasonably accurate value for the pressure in the seal chamber and to have a specified design pressure if this is greater than the seal chamber pressure (delivery, transient peak, proof test, etc.). The seal type can then be chosen from the unbalanced family of products (low pressure) or the higher pressure balanced family. Balance is described in Chapter 1, section 1.2.1. The limitation for unbalanced seals is typically quoted as 10 bar, but this is dependent on seal type and size.

The use of balanced seals is ultimately limited by the combined effects of sealed pressure and speed. This is conventionally expressed as a PV limit (see Chapter 1, section 1.3.1 for a discussion on PV factors and their limitations); PV limits differ for different designs of seals, depending on structural stiffness, and are a function of seal size.

Speed

There are three considerations. Firstly, the seal selected must be dynamically stable at the operating speed. Imperfections in the machine shafting, particularly squareness, at higher speeds will demand a greater axial tracking ability in the seal to maintain a stable fluid film.

It is common practice to select a stationary seal type for speeds greater than 4500 r/min (or about 20 m/s peripheral speed).

Secondly, where boundary lubrication conditions dominate at the seal interface, consideration should be given to face material selection of the seal.

Lastly, many face materials are severely limited in tension, thus they should not be used in an unsupported condition at higher speeds.

Temperature

The seal must either be constructed from materials which can withstand the duty temperature, or the environment must be modified to make a seal work when otherwise it would fail. The latter approach is normally expensive and introduces another degree of unreliability. Seal performance at higher temperatures is normally dependent on the secondary seal, the industry standard being about 200°C, which is the dynamic limit of elastomers.

Many seals at high temperature run with a partially mixed phase (liquid/vapour) between the faces, and face materials that will withstand this condition have to be used. An additional problem at high temperatures with seals on hydrocarbon duties is oxidation of the leaked product, leading to the formation of solid deposits (coking). To combat this, seals should be selected that will not 'hang-up', and the use of a low pressure quench (normally steam) is common practice.

Sealed fluid

The properties of the fluid have a great influence both on the selection of the primary seal and on the issue of environmental control.

Chemical properties
Materials of construction must be resistant to corrosion. In extreme cases, the seal is externally mounted so that few of the seal parts are exposed to the product, these parts being made of suitably inert materials.

Abrasion
Abrasion of the face materials can be caused by contained solids, crystallisation of unstable solutions and by-products of leakage. Particle size comparable with the fluid film thickness is particularly harmful. Film thickness is typically 0.5-2μm.

Volatility
The problem of maintaining a liquid interface with volatile products is more a matter for the seal arrangement (see Part B below) than of primary seal selection. Where there is a risk of vaporisation, the main requirement with the primary seal is to reduce the heat generation in the interface film by using balanced seals and low

friction seal face materials. In addition, the seal should have good thermal conductivity to remove heat from the critical area and reduce the ΔT requirement (see Chapter 1, section 1.3.2); this is achieved by using materials of high thermal conductivity with a narrow friction contact face. Components should also be rigid and stably balanced for mechanical and thermal effects.

Life

The life of a particular seal in a given duty is very difficult to predict. Assurances are normally based on statistical data on similar experience, rather than by analysis or test. Seals normally (but not always) fail, rather than wear out through old age; failure is almost always a function of unacceptable leakage. There are only three *real* causes of failure:

(1) wrong (or marginal) seal and/or system selection resulting from incorrect initial data or selection error.
(2) Fitting.
(3) Operating at an unspecified condition (upset, etc.).

The common failure phenomena from mis-selection or misuse are: secondary seal hang-up, overheating or dry running damage due to overpressurization, mechanical or thermal shock, abrasive wear, fatigue, elastomeric deterioration due to incompatibility or temperature (see Chapter 10).

More expensive face materials or seal designs (metal bellows seals, cartridge seals) normally give improved seal life.

Leakage

This is covered in Chapter 1, Section 1.4.1 and is again difficult to predict. It is important for the selector to work to a specified expectation and statistical historical data is again widely used.

4.3 SEAL TYPE SELECTION

Having considered all the parameters which are likely to affect the seal performance, the selector will now be in a position to select the primary seal, which comprises the seal and seat configuration and the materials of construction. Seal configuration can be categorized in many ways. The most common groups are discussed below.

4.3.1 Internally-mounted vs externally-mounted seals

Internally mounted – Advantages

– Better cooling – seal surrounded by product.
– Pressure acts to close the seal faces (pressure assisted).
– Can therefore be used at high pressure.
– Components in compression, which is preferable to tension.
– Rotating elements centrifuge particles away from seal face.
– Lower leakage due to centrifugal action.
– Most of the seal is inside machine housing, less space required outside housing.
– Seal leakage containment is simpler.

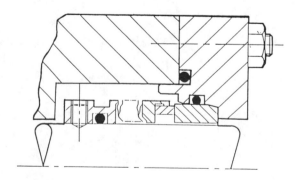

Externally mounted – Advantages

–   Easier to install.
–   Easier to inspect.
–   Minimizes components in contact with pumped fluid
    (corrosives etc.).
–   Centrifugal action can sometimes assist lubrication.

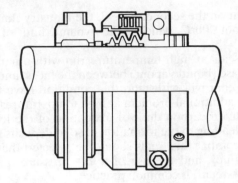

4.3.2   Rotating vs stationary floating seal

Rotating seal (Sprung member) – Advantages

–   Centrifugal action keeps particles away from flexible
    member.
–   Generally requires less axial envelope, particularly
    outside seal chamber.
–   Smaller radial section for a given shaft size.
–   Generally lower cost.

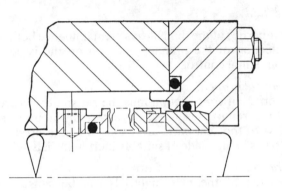

Stationary seal – Advantages

–   Capable of higher speeds.
–   Better able to cope with misalignment (particularly
    angular).
–   Less prone to clogging of leaked product if inside
    seal chamber.
–   Will accept media with higher viscosity.

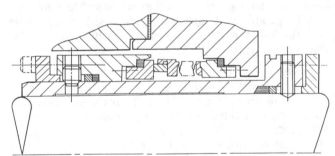

4.3.3   Balanced vs unbalanced seal

Balanced seal – Advantages

–   Capable of much higher pressures and/or speeds
    (enhanced *PV* capability).

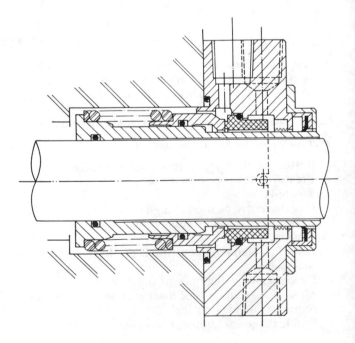

Unbalanced seal – Advantages

- Smaller envelope, particularly radial.
- No step required on shaft or sleeve.
- Lower cost.

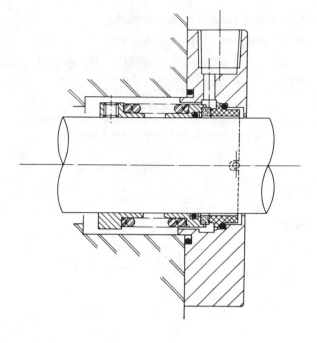

4.3.4 Non-Metal bellows seal vs pusher seal

Non-metal bellows seal – Advantages

- PTFE bellows used in very severe corrosive duties.
- Rubber bellows seals low in cost.
- Eliminate sliding packing (hang up hysteresis, sleeve wear).

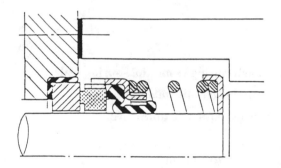

Dynamic pusher seal – Advantages

- More robust.
- Higher pressure/temperature/speed capability.
- Rubber bellows requires specially designed component in variety of materials to cope with different media.

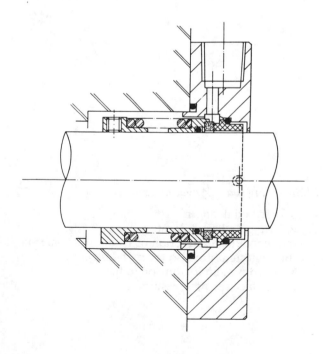

4.3.5 Metal bellows vs pusher seal

Metal bellow seal – Advantages

– Eliminates sliding packing (hanging up hysteresis, sleeve wear).
– Can be used at higher temperatures.
– Can be used at higher speeds.
– Inherently balanced without stepping shaft/sleeve.
– More compact (particularly larger sizes).

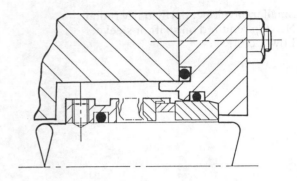

Pusher seals – Advantages

– Can be used at higher pressures.
– Less prone to fatigue failure.
– More robust.

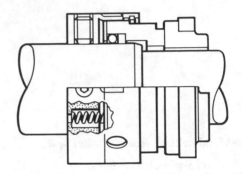

4.3.6 Single-spring vs multiple-spring seal

Single-spring seal – Advantages

– Can be used for a flexible drive.
– Larger section, more robust.
– Better protection against corrosion.
– Less prone to clogging.
– Smaller radial space.
– Low stiffness gives greater axial tolerance on fitting.

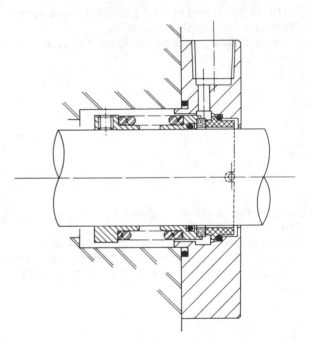

Multi-spring seal – Advantages

– Shorter axial length.
– Rotating seal can tolerate higher speeds.
– Independent of direction of rotation. (Some single-spring designs are also independent).
– More consistent loading on to face.
– Lower face spring pressure.

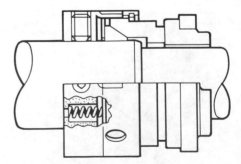

4.4 SOME SEAL CONFIGURATIONS

The following listing shows some typical seal configurations from the participating manufacturers. Their usage and capability should be checked with the manufacturer, but this listing covers almost every process or industrial primary sealing requirement. These are given purely as examples. Other seal manufacturers produce essentially similar designs; it should not be difficult to identify these by comparison with the figures.

Table 4.1 shows the normal use of each seal type in the category previously defined.

Table 4.1 Examples of typical seal configurations

Seal				Type (See Note below Table)			Illustration (Fig. 4.1)
Sealol 43	I	R	U	SS	NB	(ELASTOMER)	(a)
Crane 502	I	R	U	SS	NB	(ELASTOMER)	(b)
Crane 10T	E	R	U	SS/M	NB	(PTFE)	(c)
Crane 109B	I	R	B	M	P	(WEDGE)	(d)
Sealol 670, 676, 680	I	R	B	—	MB		(e)(f)
Crane 515E	I	R	B	—	MB		(g)
Flexibox GP1B	I	R	B	—	MB		(h)
Sealol 616	I	R	B	—	MB		(i)
Sealol 604	I	S	B	—	MB		(j)
Flexibox R20	I	R	U	SS	P	('O' RING)	(k)
Flexibox RRO	I	R	B	SS	P	('O' RING)	(l)
Flexibox FFO FFET	I	S	B	SS	P	('O' RING)	(m) (n)
Flexibox RROM	I	R	B	M	P	('O' RING)	(o)
Crane 8B	I	R	B	M	P	('O' RING)	(p)
Crane 270F	I	S	B	M	P	('O' RING)	(q)
Flexibox RREP	I	S	B	M	P	('O' RING)	(r)
Sealol HPV 5000	I	S	B	M	P	('O' RING)	(s)

Note. I = Internal   E = External
R = Rotating   S = Stationary
B = Balanced   U = Unbalanced
SS = Single-spring   M = Multi-spring
MB = Metal bellows   NB = Non-metal bellows
P = Pusher

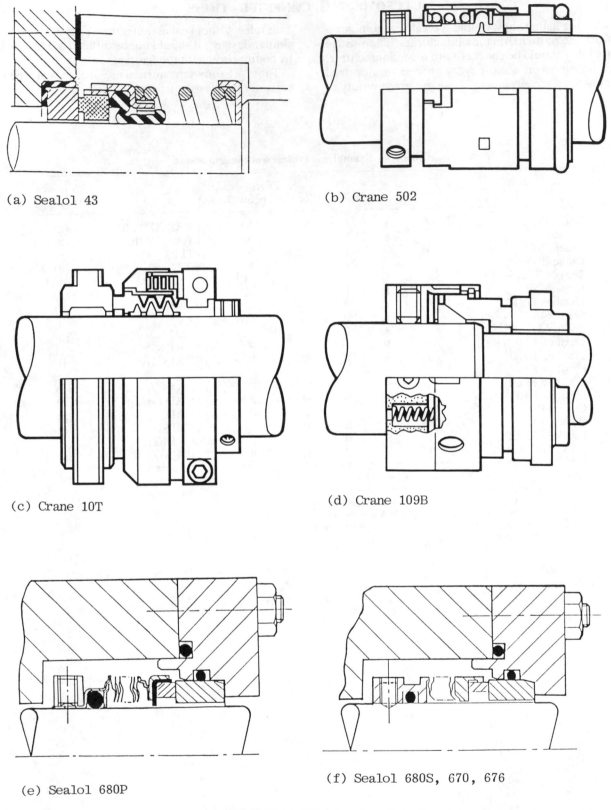

(a) Sealol 43   (b) Crane 502

(c) Crane 10T   (d) Crane 109B

(e) Sealol 680P

(f) Sealol 680S, 670, 676

Fig. 4.1   Examples of different seal configurations
(a) (b) – rubber bellows seal
(c) – PTFE bellows seal
(d) – wedge ring seal
(e) (f) – metal bellows seal – elastomeric secondary designs

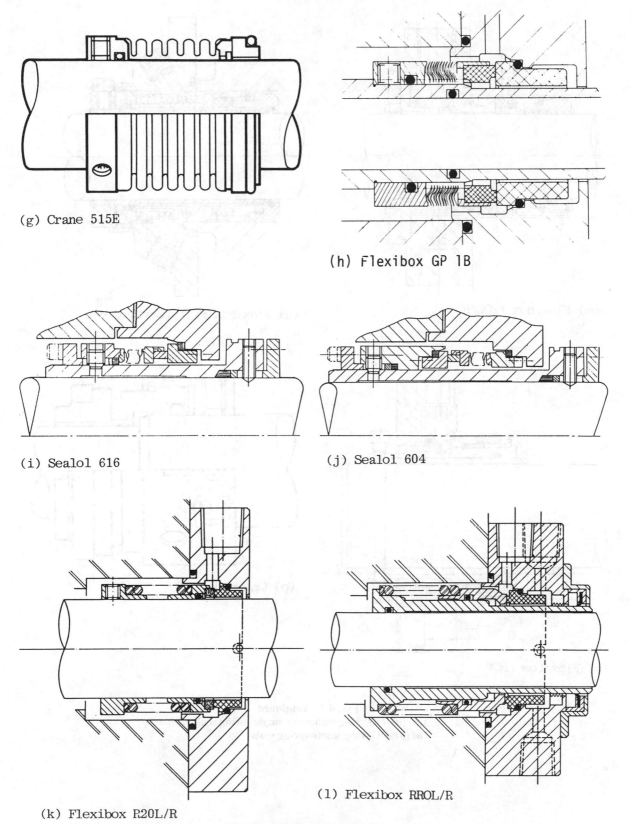

(g) Crane 515E

(h) Flexibox GP 1B

(i) Sealol 616

(j) Sealol 604

(k) Flexibox R20L/R

(l) Flexibox RROL/R

Fig. 4.1 continued
(g) (h) – metal bellows seal – elastomeric secondary designs
(i) (j) (l) – metal bellows seals – elastomer-free designs
(k) (l) – 'O' ring rotating single sring seals
continued

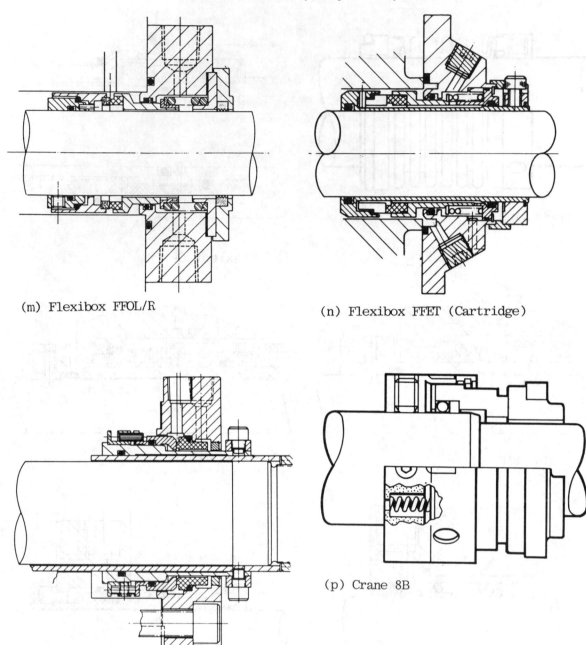

(m) Flexibox FFOL/R

(n) Flexibox FFET (Cartridge)

(o) Flexibox RROM

(p) Crane 8B

Fig. 4.1   continued
(m) (n) – 'O' ring stationary single spring seals
(o) (p) – 'O' ring multi-spring seals

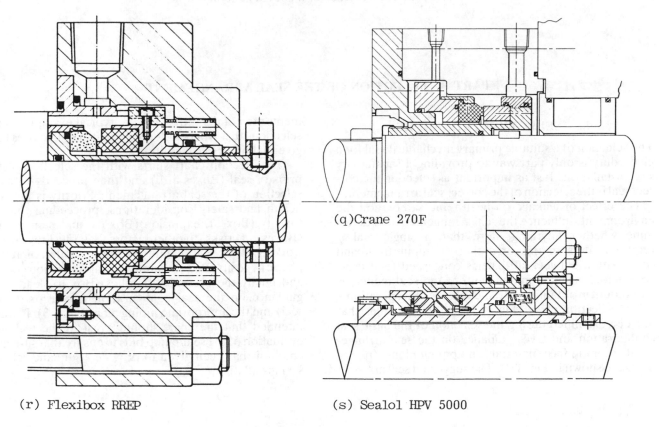

(q)Crane 270F

(r) Flexibox RREP

(s) Sealol HPV 5000

Fig. 4.1   continued
(q)(r) (s) – high duty seals

PART B   SELECTION OF THE SEAL ARRANGEMENT

4.5   SELECTION PROCEDURE

The selection of a suitable primary mechanical seal for a given duty is only part way to providing a satisfactory seal installation. Just as important as selecting the correct seal is the selection of the correct seal arrangement.

The effect of leakage on health and safety, and the environment, influence the seal arrangement and determine whether anything more than a single seal is required. It is in this area that an input from and consultation between all parties concerned (end user, contractor, machine manufacturer, and seal maker) is paramount in obtaining safe and reliable seal operation.

The selection chart (Fig. 4.2) details the areas that must be examined during the selection of the basic seal configuration and gives guidance on the seal arrangement, running face materials and piping plans, (piping plans are shown in Fig. 4.3). The suggested seal arrange-

ments are not definitive, but are typical examples of the selections that will be offered by seal manufactures for a given set of service conditions.

The selection chart starts with the selection of the primary seal (Boxes a, b) and then proceeds with the selection of the seal arrangement. Priority is given to health and safety considerations, proceeding through toxicity (Box c), explosion (Box d), and flammability (Boxes e, f) risks, to other factors that affect the correct working of the seal: liability to the formation of atmospheric deposits (Box f), liability to vapourisation (Box g) and finally the presence of solids (Box h), leading to guidance on the seal configuration (single or double seal) and the appropriate piping plan (Fig. 4.3). It is also intended that the chart should assist in the technical evaluation of the seal/pump bids to ensure that sufficient emphasis has been given to provide a safe and reliable seal installation.

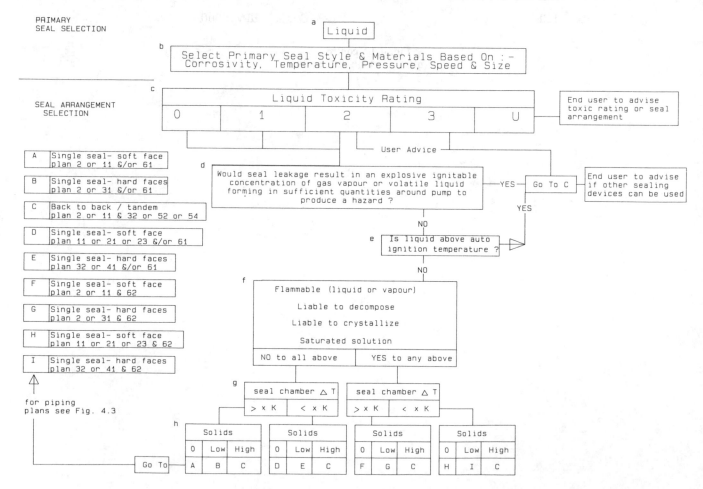

Fig. 4.2   Selection chart for seal arrangement

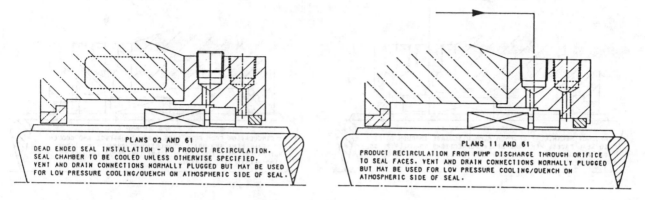

PLANS 02 AND 61
DEAD ENDED SEAL INSTALLATION - NO PRODUCT RECIRCULATION.
SEAL CHAMBER TO BE COOLED UNLESS OTHERWISE SPECIFIED.
VENT AND DRAIN CONNECTIONS NORMALLY PLUGGED BUT MAY BE USED
FOR LOW PRESSURE COOLING/QUENCH ON ATMOSPHERIC SIDE OF SEAL.

PLANS 11 AND 61
PRODUCT RECIRCULATION FROM PUMP DISCHARGE THROUGH ORIFICE
TO SEAL FACES. VENT AND DRAIN CONNECTIONS NORMALLY PLUGGED
BUT MAY BE USED FOR LOW PRESSURE COOLING/QUENCH ON
ATMOSPHERIC SIDE OF SEAL.

ARRANGEMENT A

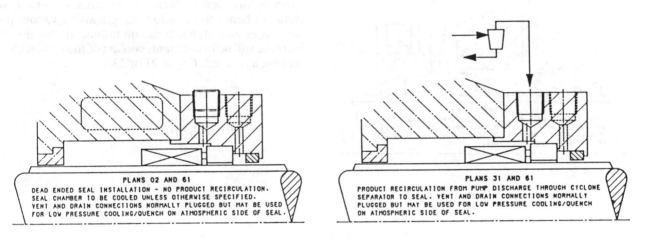

PLANS 02 AND 61
DEAD ENDED SEAL INSTALLATION - NO PRODUCT RECIRCULATION.
SEAL CHAMBER TO BE COOLED UNLESS OTHERWISE SPECIFIED.
VENT AND DRAIN CONNECTIONS NORMALLY PLUGGED BUT MAY BE USED
FOR LOW PRESSURE COOLING/QUENCH ON ATMOSPHERIC SIDE OF SEAL.

PLANS 31 AND 61
PRODUCT RECIRCULATION FROM PUMP DISCHARGE THROUGH CYCLONE
SEPARATOR TO SEAL. VENT AND DRAIN CONNECTIONS NORMALLY
PLUGGED BUT MAY BE USED FOR LOW PRESSURE COOLING/QUENCH
ON ATMOSPHERIC SIDE OF SEAL.

ARRANGEMENT B

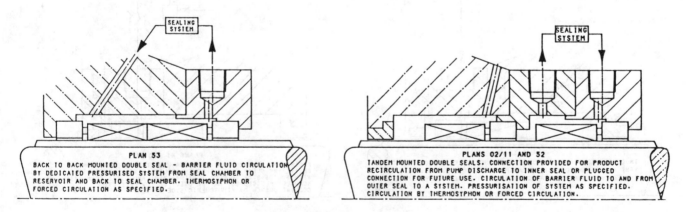

PLAN 53
BACK TO BACK MOUNTED DOUBLE SEAL - BARRIER FLUID CIRCULATION
BY DEDICATED PRESSURISED SYSTEM FROM SEAL CHAMBER TO
RESERVOIR AND BACK TO SEAL CHAMBER. THERMOSYPHON OR
FORCED CIRCULATION AS SPECIFIED.

PLANS 02/11 AND 52
TANDEM MOUNTED DOUBLE SEALS. CONNECTION PROVIDED FOR PRODUCT
RECIRCULATION FROM PUMP DISCHARGE TO INNER SEAL OR PLUGGED
CONNECTION FOR FUTURE USE. CIRCULATION OF BARRIER FLUID TO AND FROM
OUTER SEAL TO A SYSTEM. PRESSURISATION OF SYSTEM AS SPECIFIED.
CIRCULATION BY THERMOSYPHON OR FORCED CIRCULATION.

ARRANGEMENT C

**Fig. 4.3   Seal arrangements based on piping plans included in
API 610 (Drain and vent connections not shown).**

continued

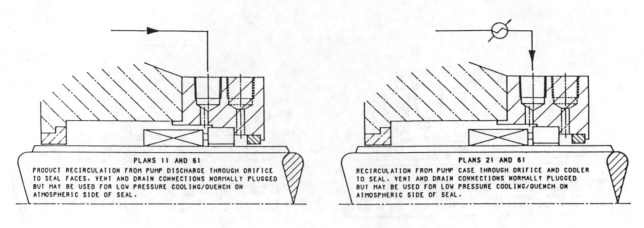

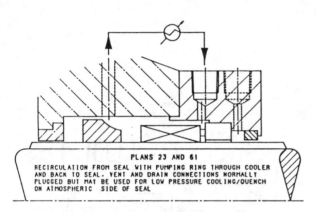

Note. The use of Plan 11 in conjunction with a Neck Bush is intended to increase the pressure in the seal chamber hence to provide a margin above vapour pressure to ensure stable face conditions. If this pressure increase will be insufficient, cooling of the product to the seal must be used. (Plans 21 or 23.)

ARRANGEMENT D

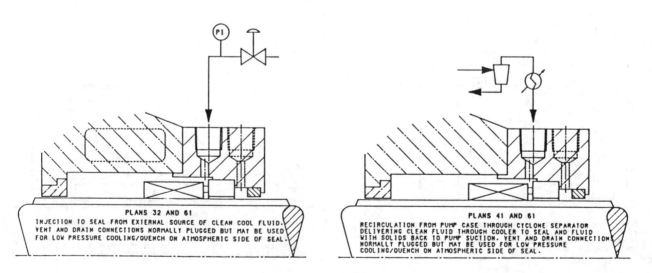

ARRANGEMENT E

Fig. 4.3   (cont.)

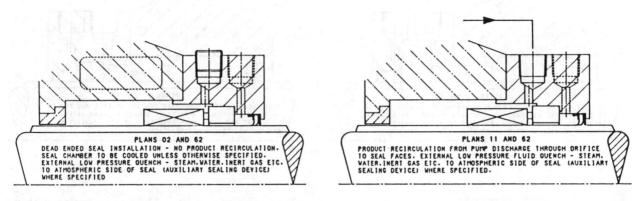

ARRANGEMENT F

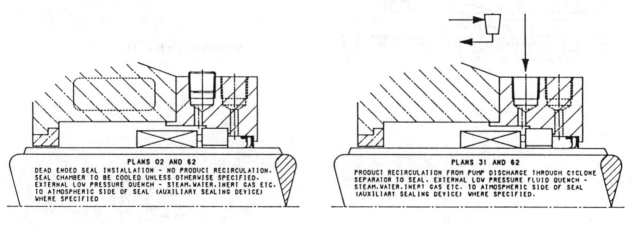

ARRANGEMENT G

Fig. 4.3 (cont.)

continued

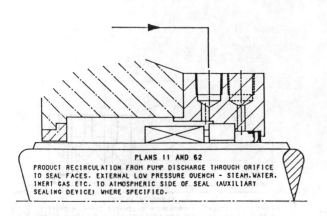

PLANS 11 AND 62
PRODUCT RECIRCULATION FROM PUMP DISCHARGE THROUGH ORIFICE
TO SEAL FACES. EXTERNAL LOW PRESSURE QUENCH - STEAM.WATER.
INERT GAS ETC. TO ATMOSPHERIC SIDE OF SEAL (AUXILIARY
SEALING DEVICE) WHERE SPECIFIED.

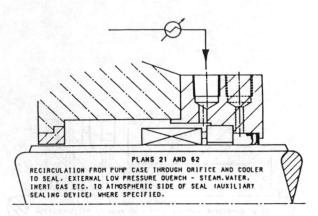

PLANS 21 AND 62
RECIRCULATION FROM PUMP CASE THROUGH ORIFICE AND COOLER
TO SEAL. EXTERNAL LOW PRESSURE QUENCH - STEAM.WATER.
INERT GAS ETC. TO ATMOSPHERIC SIDE OF SEAL (AUXILIARY
SEALING DEVICE) WHERE SPECIFIED.

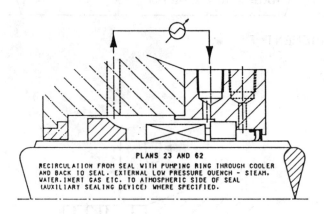

PLANS 23 AND 62
RECIRCULATION FROM SEAL WITH PUMPING RING THROUGH COOLER
AND BACK TO SEAL. EXTERNAL LOW PRESSURE QUENCH - STEAM.
WATER.INERT GAS ETC. TO ATMOSPHERIC SIDE OF SEAL
(AUXILIARY SEALING DEVICE) WHERE SPECIFIED.

Note. The use of Plan 11 in conjunction with a Neck Bush is intended to increase the pressure in the seal chamber hence to provide a margin above vapour pressure to ensure stable face conditions. If this pressure increase will be insufficient, cooling of the product to the seal must be used. (Plans 21 or 23.)

ARRANGEMENT H

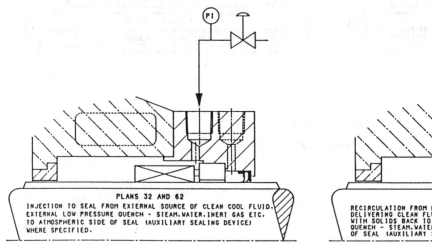

PLANS 32 AND 62
INJECTION TO SEAL FROM EXTERNAL SOURCE OF CLEAN COOL FLUID.
EXTERNAL LOW PRESSURE QUENCH - STEAM.WATER.INERT GAS ETC.
TO ATMOSPHERIC SIDE OF SEAL (AUXILIARY SEALING DEVICE)
WHERE SPECIFIED.

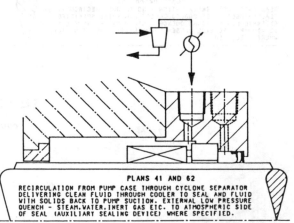

PLANS 41 AND 62
RECIRCULATION FROM PUMP CASE THROUGH CYCLONE SEPARATOR
DELIVERING CLEAN FLUID THROUGH COOLER TO SEAL AND FLUID
WITH SOLIDS BACK TO PUMP SUCTION. EXTERNAL LOW PRESSURE
QUENCH - STEAM.WATER.INERT GAS ETC. TO ATMOSPHERIC SIDE
OF SEAL (AUXILIARY SEALING DEVICE) WHERE SPECIFIED.

ARRANGEMENT I

Fig. 4.3   (cont.)

The following notes expand on the information required to answer the questions raised in the boxes in the selection chart (Fig. 4.2).

Box (a)   Liquid

Full and comprehensive details of the liquid handled together with all possible conditions of operation, i.e., off-design points, should be clearly detailed by the end user, together with all operating parameters requested on the seal data sheet. Where mixtures of liquids are pumped, details of percentage composition and vapour pressure of mixtures should be defined. Any corrosive element within the liquid should be stated (e.g., sulphur content in hydrocarbons). If details of the liquid or liquid mix are confidential, it is the responsibility of the end user, from his experience, to define suitable seal materials and advise on product hazards.

Box (b)   Select primary seal style and materials
Based on: corrosivity, temperature, pressure, speed and size.

From the details given on the pump/seal data sheet (Chapter 3, Fig. 3.1) the seal manufacturers can select a seal style and materials of construction from their product range.

4.6   SECONDARY SELECTION

Secondary selection is the consideration of 'what effect' seal leakage will have relative to site personnel, plant, and environmental safety, and how, if necessary, any leakage can be safely contained. This can be the overriding influence on the whole seal configuration and will determine whether anything more than a single seal is required.

Box c   Liquid toxicity rating

---

0   1   2   3   U

Fluids are categorised according to their toxicity, depending upon the extent of exposure or whether contact is through ingestion, inhalation, or skin absorption. This guide is based upon toxic ratings defined in the form below (Sax, 1984).

Toxicity Rating 0 – No harmful effects under normal conditions.
Toxicity Rating 1 – Short term effects which disappear once exposure removed.
Toxicity Rating 2 – May produce both short and long term effects, but normally not lethal.

Toxicity Rating 3 – May cause death or permanent injury even after short exposure to only small quantities.
Toxicity Rating U – Insufficient data available on humans.

These Toxicity Ratings are further divided into the following categories.

(1) Acute local
A single dose or exposure affecting the area of contact would affect personnel in the way described in rating 1 to U.

(2) Acute systemic
A single dose or exposure can pass through the skin, mucous membranes, or lung and be transported to the site of action within the body via the bloodstream. The action would affect personnel in the way described in the rating 1 to U.

(3) Chronic local
Repeated exposure affecting the area of contact would affect personnel in the way described in rating l to U.

(4) Chronic systemic
Repeated exposure can pass through the skin, mucous membranes or lung and be transported to the site of action within the body via the bloodstream. The action would affect personnel in the way described in the ratings 1 to U.

Each of the foregoing is further divided into how the toxic media affects the recipient, i.e., as an irritant, through ingestion, inhalation, or skin absorption.

Liquids of Toxicity Rating 0 and 1 can normally be contained with single seals.

Liquids of Toxicity Rating 2 may be contained with a single seal backed up by an auxiliary seal or a double/tandem seal, the decision being based on the end user's experience or local and national health and safety requirements.

Liquids of Toxicity Rating 3 require a tandem or double seal arrangement with a safe (non-toxic) barrier fluid. Other sealing arrangements incorporating a liquid quench may be considered, provided they meet any local or national health and safety requirements.

Glandless pumps may be considered as an alternative when handling liquids of Toxicity Rating 3.

In the case of liquids with a Toxicity Rating U, where insufficient published data is available, the end user must specify the safety requirements.

Table 4.2 indicates the seal arrangements that should be considered when pumping toxic liquids. These are minimum requirements and are given for guidance only; where local or national health and safety standards exist, they must take precedence.

Table 4.2.   Guidelines for minimum sealing arrangements with respect to liquid Toxicity Rating

Product Toxicity Rating	0		No special requirement from toxicity point of view, Single Mechanical Seal Single Mechanical Seal with Auxilliary Seal
	1		(Lip seal or Throttle Bush) with quench piping to safe area
	2	Chronic Local	Irritant – Lip Seal with Quench Piping to safe area
			Ingestion – Lip Seal or Throttle Bush with Quench Piping to safe area
		Chronic Systemic	Inhalation – Lip Seal with Quench Piping to safe area
			Skin absorption – Lip Seal with Quench Piping to safe area
		Acute Local	Irritant – Lip Seal with Quench Piping to safe area
			Skin absorption – Lip Seal with Quench Piping to safe area
	3	Chronic Local	Irritant – Lip Seal with Quench Piping to safe area
			Ingestion – Lip Seal with Quench Piping to safe area
		Chronic Systemic	Inhalation – Tandem Mechanical Seal with Plan 52 or 53 system
			Skin absorption – Tandem Mechanical Seal with Plan 52 or 53 system
		Acute Local	Irritant – Tandem/Double Mechanical Seal with Quench Piping to safe area*
			Ingestion – Tandem/Double Mechanical Seal with Quench Piping to safe area*
		Acute Systemic	Inhalation – Tandem/Double Seal with Plan 52 or 53 system
			Skin absorption – Tandem Seal with Plan 52 or 53 system
	U		Consult end user for advice on toxic hazards and safety requirements

* User to advise if other devices can be used

The above are minimum requirements and it is essential that adequate precautions are taken during inspections.

For piping plans see Figure 4.3

Box (d)   Would seal leakage result in an explosive ignitable concentration of gas vapour or volatile liquid forming in sufficient quantities around pump to produce a hazard?

If the answer to the above question is *Yes*, e.g., LPG duties, serious consideration should be given to double or tandem seals or single seals with provision for secondary containment, whatever form this takes. An alternative is to use a glandless pump.

Box (e)   Is the liquid above auto-ignition temperature?

Double or tandem seals are recommended for use on liquids pumped at or above the auto-ignition temperature. This is to prevent a fire occurring when seal leakage contacts atmosphere. Other methods, such as a clean flush to the seal, steam quench, or product cooling may be used with the user's approval.

Box (f)   Flammable (liquid or vapour)
Liable to decompose
Liable to crystallize
Saturated solution

'No' to all above	'Yes' to any above

Flammable as liquid, vapour

Any leakage that can cause a hazard (toxicity, explosion, fire) must be adequately contained from the atmosphere and disposed of. There are a number of methods depending upon degree of hazard.

The minimum containment mechanism specified in API 610 is the close clearance throttle bush. It is used in conjunction with vent and drain connections in the seal plate and allows liquid leakage to be piped to a safe place.

However, the throttle bush is not acceptable as a leakage containment component when sealing hazardous products, particularly toxic or volatile media. The conventional way to prevent primary seal leakage from entering the atmosphere is to use a double seal configuration with a compatible liquid barrier between the seals coupled to a leakage detection system (liquid and/or gas).

The seals can either be arranged in tandem or as a double (back-to-back or face-to-face) arrangement. In the case of the tandem seal, the seals face in the same direction and the chamber between the seals contains low pressure liquid. Leakage will be in the direction of pumpage to barrier. Because of this the barrier fluid can become contaminated by the pumped product; hence care should be taken with the disposal of the barrier fluid and with the selection of materials for the system. For double seals, the barrier system will be arranged with the barrier liquid pressure higher than the primary seal chamber pressure. Direction of leakage is thus from barrier to pumpage.

Between the extremes of high safety (tandem, double seals) and minimal leakage containment (throttle bush) there are a number of alternative techniques used to contain (or reduce) leakage to the atmosphere. It should be noted that at the present time there is extensive development activity in what has become known as 'secondary containment'.

There are three principle component categories to prevent or reduce leakage to atmosphere.

(1)   Clearance seals
Throttle bush
  Simple, low cost, high leakage. API 610 requires that the throttle bush be manufactured from non-sparking material.
Floating bushing
  Closer clearance than throttle bush, normally manufactured in carbon. More expensive than throttle bush (multi-component, more complex seal plate machining).
Soft/graphitic packing
  Normally unlubricated, therefore requires extreme care in assembly. Closer clearance than either throttle bush or floating bushing.

(2)   Abeyance seals
Abeyance seals are a comparatively new development in which leakage is pressure dependent. With zero or low primary seal leakage, the abeyance seal works in a similar way to the clearance seal. With deterioration of the primary seal, increased leakage passes to a safe area via a throttle which pressurises the interseal cavity. This pressure assists closure of the secondary abeyance seal until it makes full contact (zero atmospheric leakage). In this condition, seal life will be short but sufficient for safe shutdown of pump.

(3)   Contact seals
Contact seals give very low atmospheric leakage. There are two types
Lip seal
  Normally filled PTFE running against a hard coated sleeve or shaft. Normally used as a steam quench seal combined with back-up throttle (to API). Poor performance in totally unlubricated condition and at a high speed and/or high pressure.
Dry mechanical seal
  With primary seal in good condition, the secondary dry gas seal runs in dry atmospheric conditions. Dependent on the circuitry (refer to abeyance seals), the product can contain high pressure gas or liquid leaked product.

There are a number of proprietary products available from the seal manufacturers within the above categories and much current development activity.

Dry secondary containment is seen to not only reduce the cost (hardware and maintenance) of conventional liquid barrier systems, but to enhance the safety of pumps on less hazardous duties where a single mechanical seal combined with a clearance seal is the current practice.

Liable to decompose

Many products at high temperature will change form, i.e., decompose/oxidize, in contact with atmosphere. Decomposition of hydrocarbons results in the formation of carbonaceous deposits on the atmospheric side of the mechanical seal (coking); this eventually results in the failure of the seal through clogging, etc. To prevent this, low pressure steam or inert gas is injected to the atmospheric side of the seal. The steam or inert gas is normally contained by a throttle bush or auxiliary sealing device.

Liable to crystallize

Leakage of solutions containing a high percentage of dissolved solids results in the formation of crystals on the atmospheric side of the seal. The crystals eventually build up and cause seal failure through clogging, etc. A water quench is normally injected on the atmospheric side of the seal to dissolve the crystals, the water being contained by an auxiliary sealing device. Saturated solutions with crystals in suspension may cause seal face wear; in such cases mechanical seals with hard faces should be used.

Box (g)   Seal chamber ΔT

>xK	<xK

The seal chamber ΔT (product temperature margin) is the difference between the temperature of the liquid in the seal chamber and its boiling (or bubble) point at the seal chamber pressure. To ensure the existence of a stable liquid film in the seal interface, it is essential that an adequate ΔT is maintained (see Chapter 1, section 1.3.2). Typical practice is to design for a ΔT of 5–10 K minimum, but the actual value required is very dependent on the liquid concerned, the duty, seal design, and materials of construction. For this reason it is better that specific applications are discussed with the seal manufacturer. If ΔT is less than the value x agreed after such discussions, cooling to the seal or seal chamber must be applied or the pressure in the seal chamber increased to provide the required ΔT (see Chapter 1, Fig. 1.16).

Box (h)   Solids

0	Low	High

It is difficult to be specific with regard to the percentage of suspended solids that can be handled by a single mechanical seal. Seal performance in abrasive conditions is dependent on a number of factors.

(1)   Particle size.
(2)   Relative density of particles to carrier fluid.
(3)   Hardness of particles.
(4)   Concentration of particles.
(5)   Seal design.

As it is rare for sufficient information regarding particle size, relative densities, and hardness to be available, initial selections have to be made on concentrations; experience with similar applications should be taken into consideration.

Low percentage of solids

There are four main methods of sealing liquids with a low percentage of solids.

(1)   Dead ended seal cavity.
(2)   Reverse circulation.
(3)   Recirculation from pump discharge via a cyclone separator.
(4)   Clean flush.

A dead-ended seal chamber gives the simplest solution, but can be only used where no additional cooling is required (other than pump cooling jacket) to maintain an adequate ΔT. If the solids in the pumped product are sufficiently dense they tend to be centrifuged away from the back of the impeller, and thus reverse circulation (liquid flowing from the back of the impeller through the seal chamber to the pump suction) reduces significantly the solids in the seal chamber. In dead-ended or reverse-circulation seal arrangements it is preferable to use bellows seals.

Recirculating the pumped fluid from pump discharge via a cyclone separator is another method of ensuring clean fluid to the seal, the separated solids being piped to pump suction.

The final possibility is a supply of a clean flush to the seal chamber at a higher pressure than that at the back of the impeller. Flow takes place into the pump and prevents access of solids into the seal chamber. This system can only be used if dilution of the pumped fluid by the flush is acceptable; the flush rate can be restricted by using a close clearance neck bush.

High percentage of solids

Tandem or double seal arrangements are normally used when a high percentage of solids is present in the pumped fluid. If a clean flush is acceptable it should be considered. A conical seal chamber with the larger end towards the pump helps to keep the solids away from the seal.

4.7 SEAL SELECTION CONFIRMATION SHEET

A seal selection confirmation sheet is shown in Fig 4.4. This is intended to be complementary to the Seal Data Sheet given in Chapter 3, Fig. 3.1, and should be used in the technical evaluation of seal bids, or to clarify exactly the seal arrangement being offered by the seal manufacturer. Copies of the completed seal confirmation sheet and original data sheet should be returned to the pump manufacturer, plant contractor, and end user.

Seal Selection Confirmation Sheet   Date ....   Raised by ....

Ref ....

Customer ....   Ref ....

End User ....   Location ....

Plant ....   Item No.....................................

Selection

Seal Type .... Seal Size .... Seat Style ....

Unbalanced/Balanced: Internally Mounted/Externally Mounted ....

Rotating Seal/Stationary Seal   Seal Standard ....

Single Spring/Multi-Spring/Metal Bellows/Rubber Bellows/PTFE Bellow/Other ....

Single Seal/Double (Back to Back)/Double (Face to Face)/Tandem/Other ....

Secondary Device ....

API Code ....   API Plans ....

Materials of Construction

Seal Face ....

Seat ....

Secondary Seal: Seal ....   Seat ....

Spring ....   Bellows ....

Other Parts ....

Seal Plate ....   Sleeve ....

Associated Equipment

Cyclone Separator Y/N ....   Flow Controller Y/N ....

Cooler   Y/N   Type ....   Coolant ....

Coolant Inlet Temperature (C) ....   Coolant Flow ....

Filter ....   Other ....

....

Seal Order Code ....

Notes ....

....

....

Fig. 4.4.   Seal Selection Confirmation Chart

PART III

Pump considerations

PART III

Pump Operations

Chapter 5

THE EFFECT OF FITTING TOLERANCES AND PUMP VIBRATION ON MECHANICAL SEAL PERFORMANCE

P. R. Rogers

This chapter is limited to the examination of the influence on the behaviour of mechanical seals arising from the pump and the system in which it operates.

The operation of a mechanical seal is clearly influenced by the dynamic movements that take place between the rotating and stationary components. These movements result from two sources.

(1) From static misalignments arising from the squareness and concentricity of the locating and mounting surfaces, and from the alignment of the rotor with respect to the stator.

(2) From dynamic effects caused by vibration.

Quantitative limits are given in Chapter 6 on pump design.

5.1 THE EFFECT OF SEAL INSTALLATION TOLERANCES

Installation tolerances influence the seal in a number of ways and clearly some control has to be exercised over them. In order to do so we need to consider the consequences of the different effects in some detail in order to see how they vary with seal type and speed, and how they affect the choice of seal type and installation.

Mechanical seals possess the following.

(1) Limited axial movement. This is controlled by construction features and wear considerations.

(2) Free radial movement. Although radial movement produces no restoring force it is limited by the radial clearance between the shaft or shaft sleeve and the stationary seal parts.

(3) A degree of angular flexibility. This is greatly affected by seal design, particularly the form of the secondary seals and the provision of suitable clearances to permit tilting of the spring loaded flexibly-mounted components.

Axial movement (1) is provided to ensure that there is a controlled closing force between the two seal faces at all times, allowing for differential thermal movements and wear of both seal faces.

Radial movement (2) is also required to accept changes in concentricity between the bore of the seal chamber which locates the stationary components and the shaft. These pump mounted components may sometimes be located from other points than the bore of the seal chamber, but the principle remains the same. These changes in concentricity arise from forces in the casing (covered in Chapter 6, section 6.2.3), thermal movements, coupling misalignment, changes in coupling alignment, the hydraulic pump design, out of balance in the rotor assembly, excessive bearing clearance, bent shaft, and the pump operating characteristics. Some of these factors are static, others dynamic, and there is an evident relationship here also between vibration and the way in which the seal accepts radial movements.

The initial concentricity and squareness of the shaft to the reference faces in the as-built pump are primarily affected by pump component manufacture, tolerances, and the pump assembly. The reference face is normally taken to be the face of the seal chamber to which the stationary parts of the seal or its mounting plate are fitted. In addition to the pump parts, one or a number of seal parts and the mounting flange or plate are also involved in determining the accuracy with which the two seal faces are presented.

The limits given in Chapter 6 have therefore to be taken as practical starting points that allow the remaining influences, which are virtually the same as those which affect concentricity, to be accepted additionally by the seal. The freedom to accept and respond dynamically to changes to angular misalignment is termed angular flexibility.

Design plays a clear part in a seal's ability to accept angular movements. Rubber bellows seals again are normally considered to offer the most compliance to angular misalignment and, consequently, they are most suited for use on general industrial pumps where very often little thought is given to their maintenance and operation. In addition, in the bellows design there is no rubbing contact with the shaft and this eliminates the possibility of shaft damage from this cause. The degree of angular misalignment which is acceptable to the seal is speed sensitive (see Chapter 6, Fig. 6.1(b)).

The face movements induced on the sprung component can potentially induce vibration and changes in face pressure. It can be argued therefore, that the sprung mass should be kept to a minimum. Consequently, for difficult liquids, the heavier face material should be used for the stationary seal member. In this context 'difficult' is taken to mean liquids which, at the temperatures being sealed, have a tendency to produce an unstable interface film. Typical liquids would be light hydrocarbons, hot water, and many solvents.

Although not immediately obvious, one major benefit of mounting the flexible seal unit on the pump casing, rather than on the pumpshaft, is that a considerable proportion of these angular movements can be eliminated. If the counterface is fixed squarely on the shaft, then the angular movements taken up by the now stationary mounted unit are only those produced, for instance, by a rotating deflected shaft or any other event moving the impeller radially in a cyclic manner. If only one measure were possible to improve seal performance and reliability we would recommend the stationary mounting of the floating seal member.

Mechanical seals, as with all mechanical devices, are design compromises. This is not in terms of quality, but simply the match between genuine seal capability and the demands created by realistic manufacture, assembly, and operation of all the components in the pumping system. In the present state of the art it is sufficient to say that, by placing a well made mechanical seal carefully into a well designed installation which will provide a stable rotodynamic and steady hydraulic environment, the seal will provide its best performance.

From normal engineering considerations it is quite clear that, other things being equal, a seal which works true and square is going to provide optimum life. All seals work best under steady conditions, but it is rare to find them. Some seal designs, typically of the elastomeric bellows variety, tolerate considerable installation inaccuracy. However, experience shows that pusher seals and, particularly, metal bellows seals, benefit from improved accuracy of installation.

In order to improve seal performance seal makers have pursued alternative design strategies, such as complete seal cartridges and stationary mounting of the floating seal member. Cartridges remove some of the potential out-of-squareness considerations as well as the potential for inaccurate setting length. The stationary-mounted concept, which can also be incorporated into seal cartridges, mitigates some of the other factors associated with out-of-square installations. It follows, therefore, that metal bellows seals derive the maximum benefit from improving installation accuracy and design.

5.2   THE EFFECT OF VIBRATION ON A MECHANICAL SEAL

It is accepted that excessive vibration can reduce machinery life. One aspect of this has been the development of individual company and national standards which define acceptable vibration levels for various types of rotating equipment.

It is therefore important to appreciate the various sources of vibrations with their associated amplitudes and frequencies so that their influence on the performance of mechanical seals can be assessed and minimized. (See B. S. Nau (1981a) for a discussion on the effect of vibration on mechanical seal performance.)

The available evidence for the influence of vibration on mechanical seal life is statistical. It is generally accepted that there is a probability that high vibration levels will reduce seal life in pumps, but at this stage of understanding we cannot, except in broad terms, define why. For instance, as a matter of observation, the seals on the drive end of multi-stage pumps tend to suffer more than those on the non-drive end. This could be due to a variation in fluid pressure pulsation at the seal, or differences between shaft or casing vibration magnitudes and frequencies.

The effect of detailed equipment design also plays a part and it may be that different bearing stiffnesses or degrees of damping affecting each end of the pump also control the ultimate effect upon the seals and their installed life. This illustrates the subjective nature of the

available information and why a definitive answer is not yet available.

The majority of mechanical seals are fitted to centrifugal pumps, the vibration sources of which can be divided into four groups.

5.2.1   Vibration induced by the basic design of the pump

A centrifugal pump consists basically of a shaft, mounted on bearings, and carrying one or more impellers driven from an external source, such as an electric motor or steam or gas turbine. Vibrations can be induced into the pump shaft system from any or all of the following: out of balance, misalignment with the driver, bearing defects, poor manufacture of the rotor, blade frequencies, and loose components on the shaft.

Each of these sources of vibration can be identified by a discrete frequency which can readily be determined with vibration monitoring equipment.

The mechanical seal mounted on this shaft system can be subject to vibrations varying in frequency from below running speed to many thousands of cycles per second. If the frequencies happen to coincide with any natural frequencies of the components of the seal, such as the springs in a spring loaded seal, or the bellows of a bellows type seal, then fretting and fatigue damage can result. As there can be other causes, it should not be assumed that this damage is only the result of seal vibration.

It is important, therefore, to ensure that during the design of a pump all predictable resonances are designed away from exciting frequencies and heavily damped. Critical speeds should be designed to be well away from running speed, misalignments at operating temperatures minimized as these represent direct movements imposed upon the seal sprung system, the rotor components adequately balanced (see Chapter 6, section 6.2.5), and rolling bearings on vertically-mounted shafts loaded sufficiently to minimize skidding of the rolling elements.

It is worth noting that the sprung components of the mechanical seal consist of the spring, or springs, themselves, the mass of the sprung load, and a little damping from the secondary seal, such as the wedge or 'O' ring. Seals with a low sprung mass, such as those with carbon faces, together with high rate springs, will have relatively high natural frequencies. Seals with low rate springs and high sprung masses have relatively low natural frequencies, but it is extremely rare for resonance to occur even at low seal pressures. Metal bellows seals can have high or low effective spring rates and, without a sliding secondary seal, rely entirely upon film and liquid damping.

The lateral compliance of metal bellows seals is high and because of this some shaft location mechanism has to be adopted. This is often in the form of tabs or lugs in the bore of the seal face which are a relatively close, but sliding, fit on the shaft or sleeve. Metal bellows seals therefore, while not having a moving secondary seal, still require accurate installation and stable face conditions to prevent wear between the face locating lugs and the shaft surface. In unsatisfactory conditions seal hang-up can occur from this wear.

The natural frequency of most mechanical seals is well above the normal operating speed. It is particularly important with metal bellows seals to avoid axial resonances. It can also be seen that a change of face material, particularly to one as dense as tungsten carbide, can affect seal resonance considerably and is one of the reasons why the much less dense silicon carbide is preferred for the sprung face component when a hard material is required.

5.3.2 Vibration induced by the operation of the pump both under normal and upset operating conditions

Once a pump is in operation the rotor and, hence, the mechanical seal will be subject to hydraulically induced vibrations. With changing flows as the pump rate alters through the pump, the axial and radial stability of the fluid entering the impeller inlet changes. At other than best-efficiency design flow, hydraulic forces are likely to increase at the inlet of the impellers and diffusers. At low and high flows, related to the best-efficiency point, cavitation may occur. The inlet flows can also be disturbed by the liquid recirculation within the pump. The flow past the internal fine clearances will increase if wear takes place and the disturbance to the inlet flow will increase with time. At part load operation the radial forces increase on the impeller.

This mixture of operation-induced vibrations will be random in frequency and amplitude, giving rise to both mechanical vibrations of the rotor system and varying pressure conditions at the seal.

Systems should be *designed*, and pumps should be *selected* and *run*, for near peak efficiency operation and stable suction performance. When selecting a pump with a maximum diameter impeller, consideration should be given to the blade passing forces and, hence, vibration produced as a result of minimum stator to rotor tip gaps. This may apply when very quiet operation is required, for instance in hospital installations.

5.2.3 Vibration induced by distortions of the pump due to temperature and pressure changes

Temperature changes in the operating conditions of the pump can cause thermal distortions. The pump may move out of alignment with the driver. Various parts of the pump may distort in a way that induces distortion in the seal face components. Connected piping may impose significant expansion loads on the pump casing unless appropriate measures are taken to minimize it. The pump casing and rotor are affected at different rates by changing temperatures and, in general, the rotating shaft will expand or contract quicker than the casing. During warm up or cool down, or during changes in operating temperature, the seals will also be subject to low frequency movements of large magnitude on top of the rotating speed vibrations induced by the faces being out of alignment. As one face of the mechanical seal is attached to the casing, and the other to the rotor, the seal has to be capable of absorbing these differential movements.

As with temperature, pressure can also give rise to distortions in the pump casing which, in turn, impose disturbance on the mechanical seal. With pressure there is the added effect that as the seal chamber pressure changes, so the load on the seal faces also varies.

5.2.4 Vibration arising from manufacture or installation

If the faces of a mechanical seal are prevented from rotating square to each other, as has already been discussed in section 5.1, the long term seal performance may be impaired. To what degree will depend upon a number of operating factors as well as the particular seal design. With each revolution of the seal its flexible elements have to move to accommodate the out-of-squareness. As well as the effect upon the secondary seal, particularly if it is a sliding component such as an 'O' ring, or a wedge, the seal face loading will vary around the circumference of the seal faces, giving the potential for leakage on the unloaded side, and excessive wear on the loaded side.

Leakage may carry abrasives contained in the pumped liquid into the interface film and consequently accelerate the process of wear and the ultimate failure of the seal.

The effect of piping loads and coupling misalignment has already been discussed. It is an interesting observation that the seal life on steam turbine driven pumps is generally longer than that on equivalent electric motor driven units. It is assumed that this is because of the smoothness of the drive. Conversely, reciprocating engine drives can adversely affect the seal components that carry the drive to the seal faces because of the pulsating nature of the power transmission.

Seals are not normally mechanically balanced during manufacture but every effort is made in their design to minimize unbalanced forces. Nevertheless, such items as single coil springs can be a source of unbalance, particularly if they are badly made or have a 'tail' on them to provide a drive when the seal unit is driven through the spring.

Another force affecting the seal units is that induced by pumped liquid recirculation as frequently used in mechanical seal installations on pumps. To be effective this recirculating flow should be directed on to the outside surfaces of the two seal faces, but if the flow is too great the pressure exerted on the seal unit may tend to throw the flexibly-mounted face sideways, creating a similar effect to that previously described for out-of-square components.

The effect can be reduced by using tangential entry ports or even a design that distributes the recirculation flow round the seal. Where possible the flow rate should be controlled by, for instance, a close-fitting neck bush in the seal chamber, or the flow direction reversed by connecting the seal chamber to the pump suction rather than to the discharge (reverse circulation).

This last option is quite often used with vertical pumps as it also vents the top of the seal chamber, but its potential is largely ignored even on those designs of horizontal pump where sufficient pressure differential exists for it to be effective.

5.2.5 Conclusions

Standards covering all the above sources of vibration are not available because of the difficulty of isolating the degree of contribution of each source to seal perform-

ance. The seal is an integral part of the pump, which is itself tied to a complex pumping system. It is therefore important to analyse each seal installation, not just as a seal on a shaft, but as a seal in a total machine environment.

It is difficult for the seal vendors to give firm recommendations for standards of balance and installation of seal components, as at present little is known of the exact relationships between the multiple modes, frequencies, and amplitudes of vibration and seal performance.

Chapter 6

PUMP DESIGN AND MANUFACTURE FOR RELIABLE MECHANICAL SEAL OPERATION

J. K. Frew

6.1 INTRODUCTION

Since their inception, the reliability and mean time between failure of mechanical seals fitted to pumps have been inconsistent. Operational experience shows that seal performance varies relative to application, pump type and manufacture. A significant amount of research in seal face materials, hydraulic and thermal face distortion, face lubrication, seal cooling, etc., has been carried out by a number of seal makers and, while large strides in technology and seal performance have been achieved, the ultimate breakthrough which will give consistent high reliability seals has still to be made. As a contribution towards achieving this step change in operation, the pump maker should recognise those design and manufacturing aspects of pump design that affect the seal performance and, where necessary, alter pump design philosophy to ensure the pump to seal interface is idealised not only from the standpoint of the pump, but also from that of the seal.

Similarly, the seal designer should appreciate that transient and off-design pump operating conditions that can give rise to more arduous sealing design conditions do occur in practice. In certain applications, off-design pump performance may even be the norm rather than the exception. Such conditions have to be taken into account in the design and selection of the seal.

It is the seal and pump manufacturers' joint responsibility to communicate with each other and, if necessary, with the system designer to highlight and encompass all relevant design and operating parameters that are likely to affect the seal performance and reliability.

At the present state of understanding no clear relationships have been established between seal performance and life, on the one hand, and pump design, manufacturing standards, and tolerances, on the other. Thus, while intuitively stiffer pumps and tighter tolerances would be expected to give enhanced seal performance, there is currently no economic case on which such standards can be quantitatively based. The well recognised pump standards (ANSI B73.1, API 610, BS 5257, DIN 24 960, ISO 3069) do not produce any clear guidance on these points. Accepting the situation, this Chapter takes as a basis minimum recommendations that are recognised as consistent with currently achievable good engineering practice and are acceptable to the seal manufacturers. It is intended that these should provide a base against which seal performance can be judged. There is no intention of implying that the values given will provide optimum, or even for that matter acceptable, results and it is hoped that users specifying higher standards will in due course be able to provide performance data that will permit quantitative assessments of the effects of changes in the base parameters so that future recommendations can be made that will allow soundly based cost benefit judgments to be made.

6.2 PUMP DESIGN

6.2.1 Seal chamber dimensions

There still exists a requirement for direct interchangeability between soft packing and mechanical seals in a number of pump design specifications. In many cases this imposes restrictions on the seal design and limits its operating capability and flexibility. The most important design change that can be incorporated to help the seal designer is to have more radial space than is currently available in many instances at present. This would allow the seal designer scope to fit more robust seals, better capable of coping with plant excursions which can cause seal face or body damage leading to premature failure of the seal.

Table 6.1 gives seal space dimensions for standard seal designs fitted to end suction pumps. Where possible, to alleviate fitting problems and speed up seal replacement, consideration should be given to fitting cartridge seals within this space envelope. When specified by the end user, the pump axial space envelope must be capable of accommodating double or tandem seals. In all cases the pump design criteria laid out in section 6.2.4. Shaft stiffness must be adhered to.

6.2.2 Seal Chamber Pressure (Sealed Pressure)

Maximum and minimum calculated sealing pressure

The end user in conjunction with the contractor (if applicable) and the pump maker should define to the seal maker:

– the normal operating seal chamber pressure;
– the extremes (maximum and minimum) of pressure expected within the seal chamber. (These must include all system transients that may upset the sealing conditions, e.g., low pressure, cavitation of the pump.)

Typical transient or upset conditions which should be considered should include, but not necessarily be limited to, the following.

(1) Load rejection which may cause low pump sealing pressure.
(2) System component by-pass which may result in a high pump sealing pressure.
(3) System fouling.
(4) System upset due to tripping of one or all pumps operating in parallel, giving transient high pump sealing pressure.
(5) Part or overload operation of the system.
(6) Any pump overspeed conditions.

In each case, cognizance should be taken of any variance in system fluid temperature during any transients or off-load operation.

Table 6.1. Recommended minimum seal chamber dimensions (in)
(Extract from *API 610 Centrifugal pumps for general refinery service* 7th Edition. Reprinted courtesy of the American Petroleum Institute)

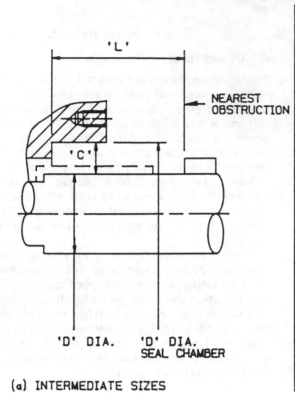

(a) INTERMEDIATE SIZES

SHAFT DIA. 'D'	SEAL CHAMBER BORE 'D'	RADIAL CLEARANCE 'C'	TOTAL LENGTH 'L'
1.062	3.125	1.001	6.625
1.183	3.250	1.031	6.625
1.313	3.375	1.031	6.625
1.437	3.500	1.031	6.625
1.562	3.625	1.031	6.875
1.687	3.750	1.031	6.875
1.812	3.875	1.031	6.875
1.937	4.000	1.031	6.875
2.062	4.125	1.031	7.125
2.187	4.250	1.031	7.125
2.312	4.625	1.156	7.250
2.437	4.750	1.156	7.250
2.563	4.875	1.156	7.500
2.687	5.000	1.156	7.750
2.812	5.125	1.156	7.750
2.937	5.250	1.156	7.750
3.062	5.375	1.156	7.750
3.187	5.750	1.282	7.750
3.313	5.875	1.282	7.750
3.437	6.000	1.282	7.875
3.563	6.125	1.282	7.875
3.687	6.250	1.282	7.875
3.813	6.375	1.282	7.875
3.938	6.500	1.282	7.875

(b) EVEN 1/8 INCH INCREMENTS

SHAFT DIA. 'D'	SEAL CHAMBER BORE 'D'	RADIAL CLEARANCE 'C'	TOTAL LENGTH 'L'
1.000	3.000	1.000	6.500
1.125	3.125	1.000	6.625
1.250	3.250	1.000	6.625
1.375	3.375	1.000	6.625
1.500	3.500	1.000	6.625
1.625	3.625	1.000	6.875
1.750	3.750	1.000	6.875
1.875	3.875	1.000	6.875
2.000	4.000	1.000	6.875
2.125	4.125	1.000	7.125
2.250	4.250	1.125	7.125
2.375	4.625	1.125	7.250
2.500	4.750	1.125	7.250
2.625	4.875	1.125	7.500
2.750	5.000	1.125	7.500
2.875	5.125	1.125	7.500
3.000	5.250	1.125	7.750
3.125	5.375	1.125	7.750
3.250	5.750	1.250	7.750
3.375	5.875	1.250	7.750
3.500	6.000	1.250	7.500
3.625	6.125	1.250	7.875
3.750	6.250	1.250	7.875
3.875	6.375	1.250	7.875
4.000	6.500	1.250	7.875

In addition, pump geometry in the region of the seal can develop localised pressure regimes that can influence the seal pressure conditions: e.g., a rotating disc or impeller shroud directly in-board of the seal will give rise to a vortex which will reduce the sealing pressure.

Pressure pulsations in seal chamber

It is accepted that large pressure pulsations can have a detrimental effect on the performance and life of a mechanical seal. Hence both the system and pump designer should recognise this and design accordingly. The maximum allowable pressure pulsations generated in the stuffing box should not exceed 5 and 10 per cent rms of the normal background sealing pressure at duty and minimum flow rate, respectively. Similarly, the seal maker should recognise such pressure fluctuations exist and design accordingly. When specified by the user the pump manufacturer must type-test pumps to prove compliance with this specification.

6.2.3 Pump casing and baseplate rigidity

Pump baseplate rigidity has an influence on the long term reliability of not only the pump, but some of its constituent components, such as mechanical seals, couplings, bearings, etc, all of which have a reduced life if casing to shaft distortion occurs. Hence it is important that realistic allowable forces and moments are specified and can be accommodated by the pump casing and its baseplate without causing undue shaft-to-casing deflection at the seal faces.

Shaft deflection, whether it be radial, axial, or angular, at the mechanical seal can have a detrimental effect on the operation and reliability of a mechanical seal. To accommodate movement the mechanical seal designer gives it radial and axial flexibility. In general, the radial flexibility is derived primarily from the flat mating faces but there is an important secondary element from the construction of the seal. Broadly speaking the bellows types of seal (sometimes defined as non-pusher designs) have better radial flexibility than those which have dynamic secondary seals, such as 'O', chevron, and wedge rings (pusher seals).

The design aspect comes into play to minimize any adverse effects of wear of the seal faces. In general, one seal ring will be made of significantly harder material than the other; it is thus important that the softer, and less wear-resistant, face should track within the face of the harder material so that steps do not develop that could interfere with radial freedom. Even so it has to be recognised that some grooving frequently develops, even on the hardest faces (carbides, ceramics), and any change in the pump rotor radial position relative to the casing forces the seal faces to take up a new running track that may cause leakage or even fracture of the softer, weaker component.

Allowable force and moments

The pump must be capable of continuous satisfactory operation when the pump branches are subjected to the forces and moments contained Table 6.2.

Table 6.2. Nozzle loadings
(Extract from *API 610, Centrifugal pumps for general refinery services*, 7th Edition.)

Force/Moment	Nominal size of nozzle flange (in)								
	2	3	4	6	8	10	12	14	16
Each top nozzle									
F_x	160	240	320	560	850	1200	1500	1600	1900
F_y	200	300	400	700	1100	1500	1800	2000	2300
F_z	130	200	260	460	700	1000	1200	1300	1500
F_r	290	430	570	1010	1560	2200	2600	2900	3300
Each side nozzle									
F_x	160	240	320	560	850	1200	1500	1600	1900
F_y	130	200	260	460	700	1000	1200	1300	1500
F_z	200	300	400	700	1100	1500	1800	2000	2300
F_r	290	430	570	1010	1560	2200	2600	2900	3300
Each end nozzle									
F_x	200	300	400	700	1100	1500	1800	2000	2300
F_y	130	200	260	460	700	1000	1200	1300	1500
F_z	160	240	320	560	850	1200	1500	1600	1900
F_r	290	430	570	1010	1560	2200	2600	2900	3300
Each nozzle									
M_x	340	700	980	1700	2600	3700	4500	4700	5400
M_y	260	530	740	1300	1900	2800	3400	3500	4000
M_z	170	350	500	870	1300	1800	2200	2300	2700
M_r	460	950	1330	2310	3500	5000	6100	6300	7200

Note. F = force (lb); M = moment (ft/lb). See *API 610 (7th Edn.)* for orientation of nozzle loads. The forces and moments given in this table are considered minimum loads and should be adjusted where the vendor has experimental or test data permitting larger values. When loads in one or more directions are significantly less than those in the table, the purchaser may request and the vendor shall then advise of load increases in the other directions that will not impair operation.

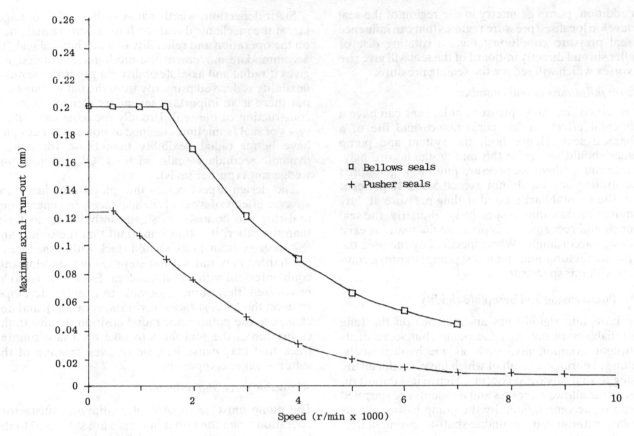

Fig. 6.1(a)   Maximum allowable axial run-out

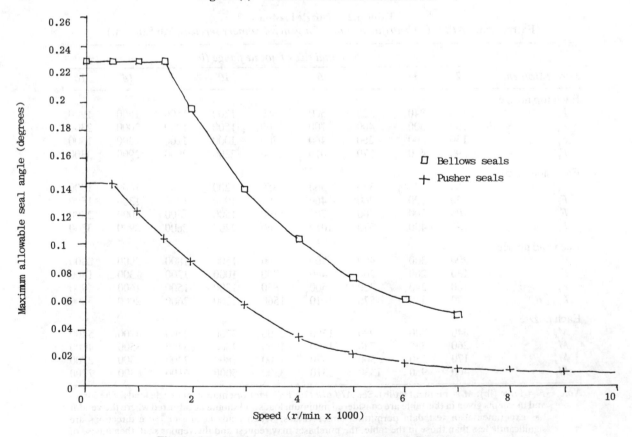

Fig. 6.1(b)   Maximum allowable static angular deflection at seal

Allowable radial and angular deflection at inway of the seal

A 'type test' record for every pump frame should be produced denoting the shaft to casing distortion at the coupling and, if practical, in way of the seal. The test should be carried out with the casing nozzles statically loaded to the values given in Table 6.2. The maximum permissible radial deflection of the rotor relative to the seal housing in way of the seal face should not exceed 0.05 mm or result in an axial run-out or angular deflection of the shaft in excess of the values shown in Figs 6.1(a) and 6.1(b) relative to the pump journal bearings.

Where the pump design prevents measurement directly at the seal, the deflection shall be measured at the coupling and the deflection, and hence angular misalignment, at the seal derived by calculation from the coupling deflection. The test should be carried out with the pump fitted to its baseplate. The pump casing to shaft displacement shall be measured in absolute terms (not relative to the baseplate). For record purposes, the pump maker shall include a schematic drawing of the test set-up, highlighting the applied forces and moments and their corresponding displacements at the shaft coupling and seal.

6.2.4   Shaft deflection at the seal

The shaft deflection is defined as the radial displacement of the pump shaft at the mechanical seal, with respect to its geometric centre line. In this section, shaft and seal sleeve run-outs, cape, and corner effects within the journal bearings have been ignored.

Allowable static deflection at seal face

The static deflection of the shaft is defined as the free shaft catenary when supported in its journal bearings, accounting for all rotor mounted weights. The free static deflection at the seal should not exceed 0.05 mm. (When running, the dynamic effects at pump internal clearances provide additional stiffness to the shaft. It is, therefore, not unreasonable that the free static deflection can be of the same order as the dynamic deflection stated below.)

Allowable dynamic deflection at the seal face

The dynamic deflection of the shaft is defined as the shaft catenary when supported at its journal bearings and deflected by rotor gravitational (static) and hydraulic (dynamic) radial forces generated at the pump impeller(s). Typical impeller radial load coefficients at zero and 100 per cent flow used when calculating the hydraulic radial loads are contained in Fig. 6.2. Where the minimum flow is greater than zero flowrate, then a linear interpolation should be used to evaluate the radial load coefficient at the specified minimum flowrate. Where more accurate coefficients exist through type or rig testing, these can be used in preference to those contained in Fig. 6.2.

Impeller radial load $(W_r) = K_r \Delta P B_2 D_t$

where
W_r = Radial load (N)
K_r = Impeller radial load coefficient (see Fig.6.2)
ΔP = Impeller generated pressure (N/mm^2)
B_2 = Impeller tip width (mm)
D_t = Impeller tip diameter (mm)

Where pumps are specified to standards which cover shaft deflection, the levels of deflection specified should be used. Such standards include API 610, ISO 5199, etc. Where there are no ruling standards, it is recommended that the design maximum dynamic deflection should not exceed 0.05 mm at the pump operating duty flow rate and 0.1 mm at minimum continuous flow rate. These levels represent current good practice. It is envisaged that these levels will give satisfactory seal life. At present there is little data available on actual shaft movement in way of the seal faces and the relationship between shaft movement and mechanical seal life.

6.2.5   Rotor balance and acceptable vibration level

Rotor balance standards

Good pump rotor mechanical balance quality will help increase seal life. A sensible and responsible approach should be taken towards rotor balance. Debatably the most destructive mode of vibration to any mechanical components within the pump emanates from out of balance forces. Hence, recognised international rotor balance quality grades should be applied:

ISO 1940   Balance quality of rigid bodies
ISO 5406   The mechanical balancing of flexible rotors
ISO 5343   Criteria for evaluating flexible rotor balance

For *rigid rotors*, the following rotor balance quality grades are recommended (A rigid rotor is defined as a rotor which has its first, wet, critical speed at least 20 per cent above the pump maximum continuous speed.):

G6.3   (G3.15/Plane of correction) for rotor operating speeds ≤ 3000 r/min, and
G2.5   (G1.25/plane of correction) for rotor operating speeds > 3000 r/min

Figure 6.3 shows the maximum residual specific imbalance corresponding to various balance quality grades, G, relative to pump rotor service speed.

The aim in balancing most *flexible rotors* should be to correct the local imbalance occurring at each component by means of balance corrections at the component itself. This will result in a rotor in which the centre of gravity of each elemental length lies on the shaft axis.

In those cases where the axial locations of the imbalances in a rotor are known, any low speed balancing technique which ensures that each imbalance is corrected in its own transverse plane will be satisfactory.

When the rotor is composed of more than two separate components that are distributed axially, it is likely that there will be more than two transverse planes of imba-

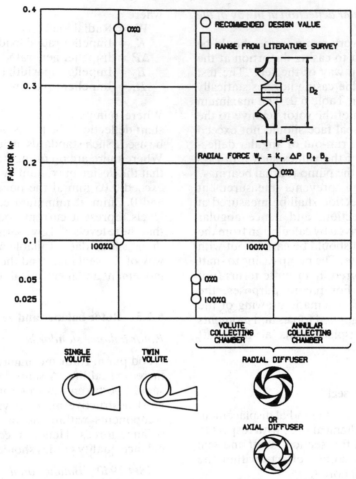

Fig. 6.2   Comparison chart for radial thrust coefficient
(Reprinted courtesy of the American Petroleum Institute)

lance. A satisfactory state of balance may however be achieved by balancing in a low speed balancing machine.

Each component and the shaft should be individually balanced as a rigid rotor to specified tolerances before assembly. In addition, the concentricities of the shaft diameters or other location features that position the individual components on the shaft should be held to close tolerances relative to the shaft axis.

The balance quality for rigid rotors is normally specified in terms of the displacement of the centre of gravity from the axis of rotation as a function of the speed. Except in a few cases, the method is at present not directly applicable to flexible rotors. In general, therefore, the balance quality for flexible rotors is specified in terms of vibration of the journals or bearing pedestals. The balance quality for flexible rotors will most probably be specified as vibrations permissible at specified measuring points or unbalance remaining in specified balancing planes.

Pump vibration levels

It is accepted that excessive vibration can reduce machine life and this has led to a variety of company, national, and international standards which define vary-

ing acceptable vibration levels for various types of rotating equipment in terms of displacement, velocity, and acceleration, the most commonly used being velocity. To date, the available evidence for the influence of vibration on mechanical seal life is statistical. It is generally accepted that high vibration levels will reduce seal life in pumps, but at this stage of understanding the effect cannot be defined quantitatively. Therefore it is difficult to analyse whether or not seal failure is entirely, partly, or not at all a function of general, or any specific mode of, machine vibration.

There are many aspects and features of the pump design, manufacture, and assembly that have a direct relation to the magnitude and type (frequency) of vibration identified on the pump. However, it should be recognised by the seal makers and plant operators that the overall magnitude of vibration will vary relative to pump operating flowrate and pump wear. For example, as a general rule of thumb the level of vibration measured on the pump bearing housing will double as the pump flowrate varies from the pump optimum design flowrate (best efficiency flowrate) down to flowrate equivalent to 10-20 per cent of the optimum design flow.

The rationale of this significant change in vibration

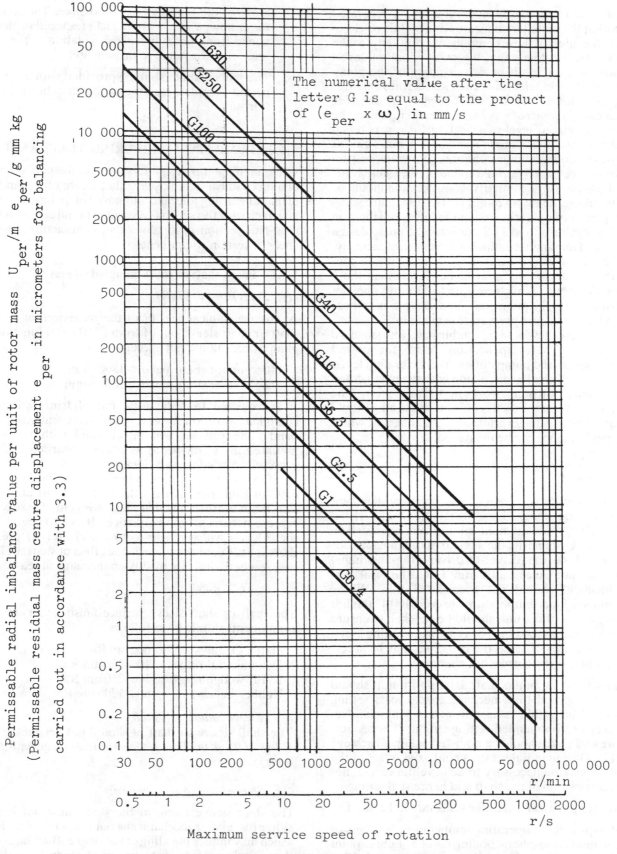

Fig. 6.3 Maximum permissable residual specific imbalance corresponding to various quality grades, *G*
(Extract from *ISO 1940/1 Balance quality of rigid bodies*)

levels is one of increased hydraulic forces generated both within the impeller and stator due to fluid re-circulation at the inlets of these components. Similarly, as the pump wears, it is likely that the pump vibration levels will increase as a result of deterioration of the pump internal and bearing clearances and loss of rotor balance.

Typically for an *ISO 3945 Criteria for assessing mechanical vibrations of machines*, Class II machine, the expected unfiltered level of vibration for a new pump should not exceed 2.8 mm/s rms and 7.1 mm/s rms at pump design and minimum continuous flowrates respectively when measured at rated speed on the pump bearing and seal housings. Similarly, consideration should be given to taking the pump out of service for refurbishing after a period of operation when the levels of vibration reach 7.1 mm/s rms and 11.2 mm/s rms at pump design flowrate and minimum continuous flowrate respectively. For operational purposes, it may be necessary to continue to operate the pump at higher levels than those highlighted above. While levels up to 18 mm/s may not be detrimental to the pump unit itself, operation at these levels is likely to cause premature seal failure.

Note. These levels are established for standard pump units. Where applications exist for critical specially engineered pump duties, likely vibration levels should be established in discussion with the pump designers and ratified on pump works performance tests. The vibration levels established on initial site testing should be established as a fingerprint for the unit against which future performance can be measured.

6.2.6 General

(1) The pump/driver radial and angular alignment should be checked on installation. The radial (parallel) misalignment should not exceed 0.05 mm TIR. Similarly, the angular misalignment (face run-out of the driver-to-driver coupling hubs) should not exceed 0.05 mm TIR. The axial misalignment should not exceed 0.1 mm. Where pumped fluid temperatures exceed 100°C the hot alignment of the unit should be checked to ensure it does not exceed the above figures. During initial set-up a cold offset at the couplings may be necessary to achieve correct hot alignment.

(2) In setting the pump shaft axial position, it should be recognised that mechanical seal axial setting should be within ±0.5 mm of the seal manufacturer's recommended setting length. If necessary, the seal plate should be adjusted axially to achieve this. During initial set up a cold axial offset of the shaft may be necessary to achieve the correct hot axial position of the shaft and hence seal setting.

(3) Where practical, cartridge seals should be fitted.

(4) Pumps with an operating temperature in excess of the fluid atmospheric boiling point or bubble point should be fitted with cooling water jackets. Cooling water jackets should be designed to surround as much of the seal cavity as is possible. The cooling water flow, and size of the jacket, should be determined from a heat balance calculation of the heat soak from the pump to the seal area.

(5) The seal chamber shall be provided with a vent to permit complete venting of the chamber prior to start-up.

6.3 PUMP MANUFACTURING TOLERANCES

While many pump design facets affect the performance of a mechanical seal, it is equally true that the standard of manufacturing tolerances imposed on pump parts that interface with the seal has a significant influence on seal life. This section rationalises the manufacturing tolerances of certain components.

6.3.1 Pump shaft or shaft-mounted sleeves

(1) Shaft run-out (TIR)

At the mechanical seal faces the maximum allowable shaft or shaft sleeve out of truth (TIR) as a function of pump speed should not exceed

0.075 mm for shaft speed ≤ 1800 r/min
0.05 mm for shaft speed > 1800 r/min

It is expected that the total out of truth will be a combination of shaft/sleeve ovality, eccentricity, and bend. The run-out will be checked with the shaft mounted in 'V' blocks at its journal bearings and the run-out recorded using a clock gauge.

On comparison with other pump standards (*API 610* 7th Edn, para 2.5.11) the level of shaft and shaft sleeve out of truth stated are less stringent. These levels represent current good practice. It is envisaged that these levels will give satisfactory seal life. At present there is little data available on the effect of shaft straightness, eccentricity, and ovality on mechanical seal life.

(2) Surface finish

The shaft or shaft sleeve surface finish under the seal components should be as follows.

Static 'O' rings	600 nm Ra
Dynamic 'O' rings	100–250 nm Ra
PTFE wedge rings	100–250 nm Ra
Rubber bellows	800–1200 nm Ra

(3) Sleeve radial fit to shaft

The shaft sleeve to shaft fit should not be slacker than H7/g6. (*BS 4500 Part 1 General tolerances and deviations*).

(4) Sleeve locking arrangements

The shaft sleeve fixing to the shaft must not impose either thermal or mechanical axial loads into the sleeve which may induce buckling of the sleeve. Buckling of the sleeve induces face distortion that leads to seal face leakage.

(5) Shaft end-play

The shaft end play on pumps fitted with anti-friction bearings should not exceed 0.08 mm with the bearing fitted. Pumps fitted with tilting pad thrust bearings will have an end play equivalent to the bearing axial clearance. If the seal is set axially with the bearing dismantled, then care must be taken to avoid over-compression of the seal.

6.3.2 Seal housing to pump facing fits

(1) Seal late run-out (face)

With the seal plate assembled to the pump casing, the run-out (squareness) on the face of the seal plate with respect to the true shaft axis should not exceed 0.08 mm/m up to a maximum of 0.1 mm TIR for pumps with an operating speed equal to or less than 1800 r/min and 0.04 mm/m up to a maximum of 0.05 mm TIR for pump shaft speed greater than 1800 r/min.

(2) Seal plate fit to pump casing

The seal plate fit to the housing shall not be slacker than H7/g6. (*BS 4500 Part 1 General tolerances and deviations*).

(3) Stuffing box bore concentricity

The stuffing box bore should be concentric with respect to the true axis of the shaft and should not exceed 0.08 mm/m up to a maximum of 0.1 mm TIR for pumps with an operating speed equal to or less than 1800 r/min and 0.04 mm/m up to a maximum of 0.05 mm TIR for pumps with operating speeds above 1800 r/min.

(4) Seal setting

The pump rotor should be set radially concentric with the stator at the mechanical seals (or local to the seals where access does not permit). The maximum allowable eccentric of rotor to stator at the seal is 0.125 mm TIR (including eccentricity associated with the seal to casing fit).

PART IV
Verification of seal design

Chapter 7

A USER'S DESIGN AUDIT CHECKLIST

B. J. Woodley

This chapter gives a checklist that is intended to assist the user with the technical evaluation of manufacturers' bids and proposal drawings. It highlights topics which, together with catalogue data, can be used to review seal selection and audit design arrangements for reliable seal operation.

Relevant topics from this checklist can be discussed with the supplier(s) using the detailed information in the earlier chapters of this Handbook. It therefore aims at promoting a dialogue between the pump manufacturer, seal manufacturer, contractor, and user to give the optimum solution for the application. In the most critical duties, for example, main oil line pump seals, such checks may need to be supplemented by seal testing, as discussed in Chapter 8.

As the checklist is based largely on one user's experience, it is unlikely to be complete. It is, however, presented in such a way that other users can modify and amplify it from their own experience. As the technology is not fully understood, some items are contentious and many are more relevant to the more difficult applications.

For ease of reference, it is subdivided into six sections.

General aspects.
Seal layout.
Seal design.
Product aspects.
Materials.
Personal notes and additions.

7.1 GENERAL ASPECTS

Topic	Checklist
Safety	Check that necessary measures have been taken. The usual adage applies, 'If in doubt, ask'.
Simplicity	The simplest seal design to meet the duty is usually the most reliable.
Design limits	Avoid operation on or near both seal and machine design limits. Seals are not fully understood and reliability usually reduces if they are taken to the design limit.
Prototypes	Avoid them. If this proves impossible, testing may be necessary (see Chapter 8).
Seal selection	Use a seal from a reputable supplier who details the basis of the selection and can provide the benefit of extensive experience.

Topic	Checklist
Supplier reference data	As when purchasing any device, a check of the track record of the supplier is worthwhile. Reputable suppliers will provide a list relevant to a particular application covering specific product, seal size, seal cavity pressure, shaft speed, etc. It is often difficult to find just one example where all the key parameters have been met simultaneously and thus judgement on the relative importance of parameters may be necessary.
Supplier/design standardization	As well as specific user/site standards, some large projects also standardize on one design for most duties to give advantages such as reduced spares holdings. To do this, the user must choose the seal supplier/ design (by competitive tender) and then specify this to the pump suppliers when inviting the latter to tender. However, it is essential to take care that the correct seal is purchased for each specific application.
Single seals	Always use, in preference to double seals, unless unavoidable for safety or ecological reasons. With difficult products, consider harder face materials, flushing, quenching, etc., before double seals.
Secondary containment	Desirable in many applications. Reasons can range from reducing water leakage from a seal entering the bearings to increasing containment of a toxic or flammable product.
	If required, consider a PTFE lip seal, two turns of Grafoil, or a non-sparking fixed throttle bushing rather than automatically using a second mechanical seal (see Chapter 4, section 4.5).
Double seals	Whether double or tandem, these normally require an intermediate sealant (barrier fluid), a method of sealant circulation, a sealant reservoir, pipework, instrumentation, etc., as well as the second seal. All can fail. Therefore, the complete system needs to be engineered for reliability (see below for some check points).
Ease of maintenance	Avoid need for special tools.
	Minimize required types/sizes of standard tools.
	Assume an adjustable wrench may well be used.
	Provide for lifting eyes, etc., on heavy seals.
	Check for adequate dismantling clearances.
	Check for details – chamfers, relief, etc. – to ease installation and dismantling.

89

7.2   SEAL LAYOUT

Topic	Checklist
Cartridge assembly	Use, whenever practical, for ease of maintenance.
	These assemblies usually contain spacers or setting washers which must be removed after the seal assembly is bolted into position and the sleeve collar is positioned in place. Typically, these are bronze setting washers on the horizontal centerline of 5 mm minimum thickness.
Shaft sleeve	Fit whenever space allows (*API610* requires a sleeve).
	Check for positive drive (by a key, drive pin, grub screws, etc.).
	To ease dismantling, check for tapped pulling holes, for relief over most of the sleeve length and for adequate sleeve thickness in accordance with good engineering practice.
	Use stainless steel (or higher alloy for corrosion resistance) with hard facing in dynamic secondary seal contact areas.
Seal chamber venting and clearances	Check design for complete (ideally automatic) venting and ample seal chamber to seal clearances to avoid running dry.
	Typically, a minimum radial clearance between the rotating member of the seal and the bore of the seal chamber of 3 mm (0.125 in) is recommended (metric designs often use 5 mm).
Anti-rotation pins	Check adequate provision for anti-rotation, particularly on high viscosity or bonding liquids.
Seal plate	Check that thickness is sufficient for pressure containment and rigidity in accordance with good engineering practice.
	Use stainless steel (or higher alloy for corrosion resistance).
	Check for adequate (e.g., 4 mm) spigot length (on inside or outside diameter).
	Confined gasket to prevent blowout may be required.
	Use bolt holes rather than slots (unless specifically required, e.g., on horizontally split casings).
	Gland stud nut access to assume use of an adjustable wrench.
	Provide for jacking screws if necessary.
Seal plate connections	Check that the size of seal plate connections is maximized to reduce the risk of damage to these connections and associated pipework (typical minimum 1/2 in NPT).
	Check for no hidden restrictions (e.g., seal plate drillings).

Check that connections are positioned correctly, not just expediently.

Check that drain connection gives complete drainage.

Check that flush/cooling is adequate – water in at bottom, out at top; steam in at top, out at bottom. Consider tangential drilling (possibly with a circumferential groove).

Check that pipework does not project into seal area causing a rub.

Check that connections are identified by stamped or cast in symbols. *API610* suggests the following symbols:

B:	barrier fluid	H:	heating
C:	cooling	Q:	quench
D:	drain	V:	vent
F:	flush		

I for in and O for out are used in conjunction with the above, e.g., a quench inlet is marked QI, or heating outlet HO.

Topic	Checklist
Double seals (barrier fluid pressure higher than product pressure)	Check for double balancing (sometimes called auto-balancing) of inside seal to prevent opening of this seal upon loss of barrier fluid pressure.

Check that seal face is positively secured in the metal bellows end piece of such seals to prevent reverse pressurization pushing the face out of the holder.

Use a clamped inboard seat to prevent even transient reverse pressurization pushing it out with consequent risk of not reseating.

Check that design has not been compromised adversely from lack of axial space.

Reverse balancing is advantageous when sealing a dirty product as this puts any solids on the outside diameter of the inboard seal and minimizes clogging. Reverse balancing is achieved by selecting the balance ratio of the inboard seal so that it can withstand pressure on the inside diameter of its face, i.e., the reverse of the normal outside diameter pressurization.

Instrumentation: typically low and high pressure alarms and/or trips and a low level alarm are required.

Use stainless steel (or better) for reservoir, pipework etc.

Check that reservoir can be easily opened for cleaning.

Check that reservoir allows any build up of gas or vapour in the barrier fluid to be vented (see Chapter 1 for an estimate of typical seal leakage rate).

Barrier fluids may become contaminated with the product in service.

Tandem seals (barrier fluid pressure lower than product pressure)

Check that design has not been compromised adversely from lack of axial space.

Metal bellows seals do not require stepped sleeves as with other balanced seals. Check that steps do not adversely compromise the sleeve thickness, result in difficulty of assembly and disassembly or an extra potential leak path (seal on second sleeve on the main sleeve).

By breaking down the pressure in two almost equal stages, identical seals can be used and the spares holding minimized.

Pumping rings and screws are notoriously inefficient circulators and experience is the only proof of pumping effectiveness. Possible approaches include:

- thermosyphon – keep pipework pressure drops low;
- tangential rather than radial connections;
- lining up the exit port with the pumping ring periphery;
- contrarotating pumping screws (where the stationary part surrounding the rotating screw has threads cut in it of the opposite hand).

Instrumentation: typically a high pressure alarm and low and high level alarms are required.

Check that reservoir can be easily opened for cleaning.

All seals leak slightly – check that system design has some method of releasing this normal leakage rather than giving false high pressure alarms (see Chapter 1 for an estimate of typical seal leakage rate).

Barrier fluids will become contaminated with the product in service.

7.3 SEAL DESIGN

Topic	Checklist
Bellows vs sliding secondary seals	Metal bellows seals avoid problems associated with sliding secondary seals and give improved performance. The use of metal bellows seals is encouraged, particularly at high temperatures, but their use is limited on pressure.
Metal bellows checks	Fatigue occurs quickly if seal is vaporizing or pump is cavitating or if stick-slip operation excites bellows torsional resonance. Check that seal is axially locked as well as positively driven (if not, balance of forces moves it along the sleeve). Corrosion rate is difficult to predict. If seal chamber flows are low the product corrosivity can become spent and seal life is longer than predicted.

Designs using large metal bellows wall thickness yield and do not return to their original length if fully compressed.

Special designs are required for duties above 10 bar ($150 \, lb/in^2$); the seal manufacturer should be consulted for advice.

Topic	Checklist
Face width	As a general rule, narrower face widths produce less heat and tend to give longer lives. Requirements vary between specific duties, but current experience suggests the following typical widths: For hydrocarbon duties: 2.5 mm (0.1 in) For aqueous duties : 4 mm (0.16 in)
Flexible face mounting	A good feature. At least one, and preferably both, seal faces should be flexibly mounted (achieved by bellows, elastomers, or springs). Particularly helpful when running two hard faces with no surface conformability together. Clamped seats are undesirable as they can be distorted when tightening the gland plate.
Seal distortion (thermal and mechanical)	Check that any shrink fit has been designed to reduce face distortion to a minimum. This includes minimizing the difference in insert/housing thermal expansion coefficients. Review with manufacturer; premium quality seals have features which increase seal life noticeably when over 200°C and 20 bar ($300 \, lb/in^2$).
Vapour locking	If liquid is sealed near its bubble point, vapour can build up adjacent to the seal faces and form a barrier which reduces face lubrication and cooling. The larger the step between the outer diameter of the primary face and the outer diameter of its parent material, the more potential for a vapour trap to form at this point.
Trapping of solids	Check for areas where centrifugal action can build up solids and hence restrict seal movement causing leakage.
Springs	In practice, one large spring is usually more reliable than many smaller springs, except for high speeds ($D_m n > 6500 \, mm \times rev/s$). Spring material should be 316 stainless steel (or higher alloy for corrosion resistance).

7.4 PRODUCT ASPECTS

Topic	Checklist
Vaporization	To avoid vaporization, an adequate product temperature margin is required (see Chapter 1, Fig. 1.12). The requirement

	varies with seal layout, design and materials.
Viscosity	If a centrifugal pump can pump it, a seal can seal it.
	Check for possible product solidification on pump cooldown, etc.
Density	Check relative density of product and solid contaminants. If solids are less dense they will not centrifuge (see section 7.3, Trapping of solids).
Abrasives/solid contaminants	Common cause of mid-life seal failure. The seal faces act as a filter and as the liquid enters the faces, the filter cake left behind clogs the entrance to the fluid film and starves the seal faces of lubrication. This causes deep scoring and possibly heat checking. (See Chapter 10 on Failure diagnosis: use of flushing, etc.)
Crystallization	Check if the fluid being sealed forms a residue when it evaporates on the seal outlet. This can glue the faces together when shutdown, or cause mechanical damage when running. (See Chapter 10 on Failure diagnosis: use of quench, etc.)
Toxicity	Although the increasing legislation in many countries has no effect on seal performance, the environmental implications have a major effect on seal layout (see Chapter 4, section 4.5).

7.5 MATERIALS

Topic	*Checklist*
PV values	Designing with these does not guarantee success. Useful in a relative, but not an absolute sense.
Low friction face materials	This is desirable as it gives lower heat generation between seal faces and reduced damage if a seal runs dry.
	Below 260°C, the lowest friction combination currently available is carbon–graphite against silicon carbide.
Carbon	Care is required as carbon grades and quality are very variable (over 100 different grades available). Key properties are hardness, Young's modulus, chemical resistance, resin and other filler content, thermal conductivity and temperature limit.
	Discussion with manufacturer is required to check the suitability of a grade for a particular duty.
Silicon carbide	So successful that many users standardize on its use. Reaction bonded silicon carbide (e.g., Refel SiC) is usually preferred, but some recently introduced grades give

	improved performance in certain applications.
	Take care on source and quality of supplies.
Tungsten carbide	Preferred form is a solid insert with nickel binder. The nickel binder has improved corrosion resistance over a cobalt binder.
Alumina	For maximum chemical resistance, use the 99.5 per cent purity grade. Useful in aqueous and mineral acid duties, otherwise avoid.
Stellite	Relatively low initial cost, but not recommended for long life and reliability – prone to heat checking and grooving.
Coatings	Beware of cracking, blistering and peeling of hard coatings; use solid faces for maximum reliability.
Elastomers	One of the greatest causes of seal life unpredictability. They age, harden, soften, relax, extrude, swell, shrink, melt, and dissolve. For longest life, minimize usage particularly in sliding secondary seal positions (see above, section 7.3, Bellows versus sliding secondary seals).
	Large number of manufacturers and grades – apparently similar materials with the same generic name, e.g., nitrile, have dissimilar properties.
	Ethylene propylene rubber can fail because oil or grease was used to ease assembly and similarly silicone rubber is attacked if silicone oil is used during seal installation (see Chapter 9, section 9.3).
PTFE	Purpose designed metal/PTFE constructions give optimum results as they are more resistant to thermal cycling than solid PTFE.
	Solid PTFE parts such as 'O' rings lack elasticity and, hence, are prone to (a) damage during assembly, (b) relaxation under load and extrusion through small clearances.
	For long term reliability, avoid PTFE jacketed 'O' rings (with a core of elastomer to provide resilience). Cracking, etc., of the thin jacket is followed by rapid elastomer attack on corrosive duties.
Moulded exfoliated graphite foil	Typical proprietary names are Grafoil and Graftite.
	Good for high temperature but has to be handled with care during installation.
	It absorbs water and so, if it is exposed to water and then heated, the seal face can be pushed out of true. Check that the seal ring is a shrink fit if this could occur in a particular application.

Hardfacing   Colmonoy/Metco type treatments on sleeves, etc., reduce wear under dynamic secondary seals, under secondary containment PTFE lip seals or Grafoil and under anti-vibration lugs of metal bellows seals.

         Check that hardfacing does not extend to grub screw area as drive grip may be impaired.

7.6 PERSONAL NOTES AND ADDITIONS

Topic	*Checklist*

Chapter 8

SEAL TESTING AND VERIFICATION OF SEAL SELECTION (CRITICAL DUTIES)

P. J. Dolan

8.1 INTRODUCTION

8.1.1 The present situation

Until now most seals have been subjected to static pressure tests which aim to ensure that mechanical strength and assembly are satisfactory. Dynamic testing, if performed at all, has usually been performed on cold water or lubricating oil. In practice it has been found that seals can pass these tests but fail to give satisfactory performance in service. This is attributed to the fact that seal selection was incorrect and the tests do not adequately represent service conditions.

It is the aim of these guidelines to suggest ways of testing seals which will give a much higher probability that seals which pass these tests will give satisfactory service.

8.1.2 The purpose of testing

There are two purposes in testing seals.

(1) To ensure that the seal has been correctly manufactured and assembled.
(2) To verify that the seal has been correctly selected. This means that the seal is suitable for the operating conditions and will give acceptable leakage and life.

Tests aimed at the second objective would be performed on a representative sample seal, i.e., a type test.

8.1.3 Degree of testing

The purchaser should decide at the time of ordering what degree of testing he requires, taking into account the seal vendor's normal quality control procedures and the additional costs which may be incurred. Depending upon the scope of testing, the costs of verifying seal selection may be considerable.

It is suggested that for the following types of duties verification of seal selection is not necessary.

(1) Duties where there is good reference experience.
(2) Duties where the consequences of failure would not be serious.

For critical duties, where the consequences of failure would be serious and the reference experience is inadequate, some verification of seal selection may be desirable. For example, such seals could be main oil line pump seals or power station boiler feed pump seals.

The degree of testing required depends upon the criticality of the duty and the relevance of the vendor's experience and works tests. It is not intended to suggest that verification of seal selection should be performed for all seals or that all features should necessarily be checked for a seal where some checking is felt to be justified.

8.2 TYPE OF TEST

8.2.1 Check on manufacture and assembly (seal vendor's works)

A simple hydrostatic pressure test will be sufficient. Since dismantling the seal would defeat the purpose of the test only cartridge seals can usefully be tested before assembly into the pump.

8.2.2 Verification of seal selection

There are many features of design to be considered in seal selection. These are discussed in Chapter 4.

8.3 FEATURES OF DESIGN TO BE VERIFIED

The checking of seal design features is discussed below. The tests involved are summarized in Table 8.1.

8.3.1 Material selection (laboratory)

Seal materials can normally be selected without the aid of experimental tests. If it is felt that experimental work is required, this will usually take the form of immersing samples of material in the fluid with which it will be wetted. Service conditions of temperature and pressure should be matched.

Materials which rely on forming a passive surface film for their corrosion resistance can suffer rapid corrosion under dynamic conditions. Thus dynamic corrosion/wear testing may be necessary if no relevant experience is available with the particular face material combination in the fluid to be sealed.

Elastomers in contact with hydrocarbons at high pressure may suffer explosive decompression if pressure is suddenly reduced. If large and rapid pressure swings are expected, tests can be carried out in equipment which simulates such transients.

8.3.2 Mechanical strength and correct assembly (seal vendor's works)

These are normally tested by submitting the seal to a hydrostatic pressure test. Such a test will not verify the strength of the drive to the rotating face nor the strength of the anti-rotation elements which secure the stationary face.

8.3.3 Lubrication regime (seal vendor's works)

For a seal to operate satisfactorily it is necessary to establish a suitable lubrication regime between the faces. Theoretical means of predicting the regime between seal faces are highly complex and not suited to use by purchasers. They are also not fully validated. It is therefore necessary to resort to testing by methods discussed below (see section 8.4.2).

8.3.4 Product temperature margin (seal vendor's works)

In operation the relative motion of the seal faces generates frictional heat. This causes the face temperature to rise until the temperature difference with the surroundings allows the heat to be dissipated and a state of equilibrium is reached. If the temperature rise is too great, the liquid between the faces vaporizes. Depending on the degree of vaporization, this can be very destructive and lead to rapid failure of the seal faces.

Methods of predicting the temperature margin required are not precise. It is therefore necessary to resort to testing by methods discussed below (see section 8.4.2).

8.3.5 Resistance to swash and vibration (seal vendor's works)

All seals experience some degree of swash and vibration in service. Normal seal test rigs are relatively free of both. If it is felt that swash and vibration pose a particular problem, some test rigs offer the possibility of building in these features. Such tests should be discussed with the seal vendor.

8.3.6 Tolerance of transients (seal vendor's works)

It is generally accepted that seals work best when operating conditions are kept constant. Changes in conditions are likely to cause relative rotation (coning) of the seal faces as the result of distortions resulting from temperature or pressure. Greater amplitude of change or more rapid change are both likely to make the effects worse.

8.3.7 Resistance to abrasion (service experience)

In general, realistic abrasion tests are extremely difficult to arrange other than in service. This is because of the following.

(1) The abrasive tends to settle in quiet areas of the test rig.
(2) The abrasive which circulates in the rig may get blunted and become less aggressive.
(3) The abrasive may suffer comminution. The change in grain size will change the aggressiveness of the environment.

For these reasons it is unlikely that useful tests of abrasion resistance can be performed other than in service.

8.3.8 Resistance to solids deposition (service experience)

If the liquid contains solids in solution, these may be deposited by crystallization and/or evaporation between the seal faces where they will cause rapid wear. They may also deposit on the atmospheric side of the seal where they can prevent axial movement of seal components. Deposits can also act as a cutting tool to machine rotating parts (usually the shaft sleeve).

The risk of solids deposition can be assessed from a consideration of the pumped liquid. The remedies are well known and it is not normally considered necessary to conduct special tests.

8.4 DYNAMIC TESTS (TYPE TESTS)

The aim of these tests is to verify that the seal is operating with a suitable lubrication regime and product temperature margin. A further requirement may be to verify the seal's tolerance to transient operating conditions, swash and vibration.

Since the tests are intended to verify design a type test is sufficient. In this context 'type' means a seal of the identical design, size and materials of construction. Testing of all seals in an order is not required.

8.4.1 Class 1 test

The ideal test is to operate the seal at the specified seal chamber conditions on the specified pumped liquid. This is known as a Class 1 test. Unfortunately this is rarely possible because of the following.

(1) Difficulty of providing contract fluid in vendor's works.
(2) Safety considerations of handling flammable and/or toxic liquids in vendor's works.

Thus Class 1 tests can usually only be performed where the service fluid is water or lubricating oil.

8.4.2 Class 2 test

If a Class 1 test cannot be performed it is necessary to perform tests on another fluid at conditions which are equivalent to the specified duty conditions. Such a test is known as a Class 2 test. It is not always possible to test both lubrication regime and product temperature margin simultaneously. A two stage test procedure, in which they are tested separately, may therefore be necessary.

Lubrication regime

The lubrication regime between the seal faces has similarities with that in a thrust bearing and may be characterized by a non-dimensional Duty Parameter G (Chapter 1, section 1.2.3) which is analogous to the Sommerfeld Number of bearing theory. The Duty Parameter is defined by

$$G = \eta \frac{Vb}{F_t}$$

Where the symbols are as defined in Appendix 2.

If G is too small, an inadequate fluid film will be formed and increased wear will result.

If G is too large, the fluid film will be too thick and increased leakage will result.

It is therefore necessary when testing a seal to arrange that the value of G on the test bed is similar to that under the specified service conditions. Change of pressure and/or temperature can cause differential rotation (coning) of the seal faces and thereby change the hydro-

Table 8.1.   Summary of seal tests

Paragraph	Feature verified	Test	Site
8.1.2(1)	Correct manufacture and assembly	Hydrostatic	Seal vendor's works
8.3.1	Material selection	Static/dynamic Explosive decompression	Laboratory
8.3.2	Mechanical strength and assembly	Hydrostatic	Seal vendor's works
8.3.3	Lubrication regime	Rig test (Section 8.4.2)	Seal vendor's works
8.3.4	Temperature margin	Rig test (Section 8.4.2)	Seal vendor's works
8.3.5	Resistance to swash and vibration	Rig test (Section 8.4.2)	Seal vendor's works
8.3.6	Tolerance of transients	Rig test (Section 8.4.2)	Seal vendor's works
8.3.7	Resistance to abrasion	No test	In service
8.3.8	Resistance to solids deposition	No test	In service

static contribution to load support. It is therefore necessary to test seals at conditions as near as possible to the service conditions of pressure, temperature, speed, and viscosity. The speed can be adjusted to make correction for unavoidable small differences between test and service conditions and achieve the correct value of G.

To assess wear rate, tests will probably need to extend over at least 100 hours with measurement of faces before and after test.

Some face materials, such as ceramic, PTFE, and stellite, are wetted by liquids of low surface tension, such as hydrocarbons, but not by liquids of higher surface tension, such as water. This can lead to failure to form a fluid film between the faces. The seal will then fail a test on water, although it would be satisfactory on hydrocarbon. If such face materials are to be tested on water the surface tension should be modified to that of the service liquid by the addition of surface active agents.

Temperature margin

The temperature margin is defined as the difference between seal chamber temperature and the boiling point of the liquid at seal chamber pressure. The phrase 'product temperature margin' is often shortened to ΔT.

In order to test for ΔT, it is necessary to arrange that heat generation on test is the same as under specified conditions. Thus speed and pressure should be the same as specified. The test temperature should be adjusted so that ΔT available on test is not greater than at the specified duty conditions.

Liquid properties (viscosity, specific heat and thermal conductivity) influence ΔT required, but at present it is thought the effects are relatively small and can be ignored.

The test is regarded as satisfactory if no 'popping' of seal faces occurs.

Swash

To test for tolerance of swash requires the ability to mount the rotating seal face not quite perpendicular to the shaft. Most seal manufacturers have rigs capable of such a test. The pump manufacturer should specify the amount of swash expected. See Chapter 6, section 6.3 for acceptable limits.

Transients

The nature of transients is so variable that no standardized test can be recommended. In setting up such a test one should attempt to match the magnitude and rapidity of change which is anticipated in the service conditions. Such changes should include pressure swings, temperature swings, change of speed, and stop-start operation.

Eccentricity

The effect of eccentricity depends upon which face is mounted eccentrically. If the rotating face does not rotate about its geometric centre, this will cause leakage. Eccentric mounting of the stationary face may be significant, depending upon the relative widths of the faces and the degree of eccentricity.

The degree of eccentricity is determined by pump and seal manufacturers' tolerances and shaft deflections. See Chapter 6, section 6.3 for acceptable limits.

Most seal manufacturers have rigs capable of testing seals under conditions of eccentric mounting.

8.5   ACCEPTABLE PERFORMANCE

For satisfactory operation all seals have a fluid film present between the faces. Therefore all seals leak when operating, although sometimes this leakage is in the vapour phase giving the impression of a leak free seal.

A guide to normal leakage rates is given in Chapter 1, section 1.4.1. If increased leakage is acceptable, it may be possible to increase film thickness and reduce wear in the seal. This must be agreed between vendor and purchaser.

Wear of seal faces can be measured and extrapolated to give an estimate of seal life. There is evidence that the wear rate in correctly operating seals decreases with time. Thus extrapolation of wear rate to estimate seal life is likely to be pessimistic.

Popping of the seal faces leads to rapid failure. No popping is acceptable during the ΔT test.

Seals should not be stripped for examination on com-pletion of a satisfactory hydrostatic pressure test. On completion of dynamic type tests the seal should be stripped and examined. All components should be in good condition.

PART V

Practical considerations in using mechanical seals

Chapter 9

INSTALLATION AND OPERATION
B. J. Woodley

Mechanical seals are relatively precise items of equipment. For reliable operation, seals require the correct working environment, which in turn demands good engineering practice in plant operation and maintenance. This may require appropriate training.

These practical aspects of installation and operation have a significant effect on seal life expectancy. They are also within the control of the seal user, who directly benefits from any improvement.

The topic is divided as follows.

9.1   Seal handling and inspection.
9.2   Pre-installation machine checks.
9.3   Seal installation.
9.4   Seal removal.
9.5   Seal modification.
9.6   Commissioning and operation.

A tabular format is used for ease of reference.

9.1   SEAL HANDLING AND INSPECTION

These notes cover pre-installation checks applicable to the mechanical seal itself. They are aimed at providing a measure of the care and attention required by seals throughout the fitting process. Spare parts storage is also referred to in this section, as it has a direct effect on an apparently new seal.

Topic	Description	Recommendations/remarks
Handling and fitting instructions	Specific instructions on: arrangement drawings; data sheets; handbooks, etc.	Specific instructions overrule general recommendation. If in doubt, vendors and other consulting services are available for advice.
Care to avoid mechanical damage	Many seals are brittle and/or fragile. If dropped, non-metallic parts may shatter.	Protect parts from damage wherever possible. Avoid placing a sealing face down on any surface unless it is protected by a clean cloth or similar.
Care to avoid other damage	Some parts are prone to attack by common liquids.	For example, ethylene propylene rubber is attacked by mineral oil and silicone rubber is attacked by silicone oil. See section 9.3 for recommended lubricants to be used in fitting.
	Heat to remove an adjacent part.	If unavoidable, then seal areas may need to be cooled locally.
Packaging	Usually skin packaged and boxed.	If a visual pre-check is made, it is usually best to return the seal to its original container. Some packaging requires a knife to open. Elastomers and seal faces should be checked for cuts and damage after packaging has been removed.
	Label on boxes.	A useful cross-reference to order number, part description, etc., which gives a check that the seal is correct for the duty.

Topic	Description	Recommendations/remarks
Arrangement drawing	Sectional view of specific seal installation.	Most vendors will supply if requested (ideally on initial enquiry). This gives not only key dimensions for later checks, but also specific seal maker's recommendations, which can be transferred to maintenance procedures.
Maintenance procedures and training	Formal written procedures which are increasingly common, even in smaller companies, when safety is involved. With the increasing complexity of maintenance procedures, training of craftsmen may be required.	Transfer of all technical data and specific seal maker's recommendations to relevant maintenance procedure for future use. Note especially any safety/ toxicity/industrial hygiene issues.
Fitting instructions	Details supplied with the vast majority of seals.	Read them. Follow them.
Physical checks of the mechanical seal. (Some more precise checks are given in Chapter 10.)	Seal rotary and stationary units.	(1) Check for physical damage. (2) Ensure drive pins and/or spring pins are free in the pin holes or slots. (3) Check set screws are free in the threads. Note: replace set screws after each use. (4) If there is a concern with a metal bellows type seal, a liquid paraffin test can be used (see Chapter 10, section 10.6).
	Both seal faces.	Visually check that there are no nicks or scratches. Imperfections of any kind on either of these faces can cause seal leakage.
	Gasket thickness.	Where possible, check thickness against drawing dimensions as the incorrect thickness can affect seal setting and hence compression.
	Spring handing on single spring seals.	The spring should be installed so that in operation the traction force from the friction at the faces acts to tighten the spring. *Note*: some single spring seals are bidirectional.

L.H. SPRING | L.H. SPRING | R.H. SPRING

INTERNALLY MOUNTED SEALS

X = LAPPED FACE OF ROTARY SEAL RING

R.H. SPRING

EXTERNALLY MOUNTED SEAL

COIL END

ANTI-CLOCKWISE ROTATION

Looking toward the sealed fluid

LEFT HAND

LEFT HAND SPRING

Topic	Description	Recommendations/remarks
Spare parts storage	Storage of seal assemblies or individual seal parts prior to installation.	If requested, seal manufacturers will supply a spares holding recommendation. Being relatively delicate, seals are best stored in the supplier's protective packaging. The stores area itself needs to be clean, dry, and adequately warm and ventilated. The user can choose between holding complete seals or individual seal parts to build up to complete seals. Although the latter is normally more economic (as the numbers of identical parts used in different seals can be reduced), it does require a more formal and ordered stores system.

9.2 PRE-INSTALLATION MACHINE CHECKS

Although precise correlations between seal performance and the mechanical condition of the machine in which it is installed have not been established, it is recognised that seal life is adversely affected by misalignment and vibration. The notes below give limits, readily obtainable with good engineering practice, that should create the conditions necessary to give the best chance of a long and cost-effective seal life (see Chapters 5 and 6).

Topic	Description	Recommendations/remarks
Shaft straightness	Total out of true between centres, i.e., maximum total indicator reading (TIR) of the shaft or shaft sleeve at the location of the mechanical seal faces. 	0.1 mm (0.004 in) maximum for shaft speed $\leqslant$1800 r/min. 0.05 mm (0.002 in) maximum for shaft speed >1800 r/min. If close to these limits, then very accurate bearing location is required (see shaft run out below).
Rotational balance	Of rotary shaft assembly (including impeller, etc.). 	At least of satisfactory standard to *VDI 2060* or *ISO 1940* quality grade levels: G6.3 (G3.15/plane of correction) for rotor speeds, $\leqslant$3000 r/min. G2.5 (G1.25/plane of correction) for rotor speeds >3000 r/min. Improving the balance quality will increase seal life only if other mechanical standards are similarly upgraded.
Shaft run out	Run out (TIR) at the seal with the shaft located in the pump bearings (where the design allows). 	0.1 mm (0.004 in) maximum for shaft speeds $\leqslant$1800 r/min. 0.05 mm (0.002 in) maximum for shaft speeds >1800 r/min. Lightly lifting the shaft may show a greater reading than shaft run out indicating bearing wear (see below).

Topic	Description	Recommendations/remarks
Shaft bearing clearances	Radial movement at the seal.	0.08 mm (0.003 in) maximum for rolling element bearings. For plain bearings, radial movement not to exceed the maximum bearing clearance specified by the machine manufacturer.
	Axial shaft movement or end float.	0.08 mm (0.003 in) maximum. Where axial movements exceed this by design, e.g., with tilting pad thrust bearings, the seal design or seal installation procedure must prevent over and under compression. Normally, hydraulic force will locate a shaft in one position, but abnormal pumping conditions or stop/start operations may well reduce seal life (e.g., by causing fretting at dynamic secondary seal contact).
Shaft/sleeve diameter	Shaft size under the seal (especially in the secondary seal position).	±0.05 mm (0.002 in) of nominal diameter. Broadly independent of type of seal.
	Shaft maximum ovality/eccentricity under the seal (especially in the secondary seal position).	Dependent on the type of secondary seal. Static 'O' ring (i.e., metal bellows seal): ±0.025 mm (0.001 in). Dynamic 'O' ring: ±0.025 mm (0.001 in). PTFE wedge ring: ±0.025 mm (0.001 in). Elastomer/rubber bellows: ±0.05 mm (0.002 in).
Shaft/sleeve surface finishing	Surface finish under the seal (especially in secondary seal position); see also later comments about sleeve hard facing.	Dependant on the type of secondary seal: Static 'O' ring (i.e., metal bellows seal): 600 nm Ra (25 μ in). Dynamic 'O' ring: 100/250 nm Ra (4/10 μ in). PTFE wedge ring: 100/250 nm Ra (4/10 μ in). All above to be ground and polished free from machining marks. Elastomer/rubber bellows: 800/1200 nm Ra (30/50 μ in). Fine machined: a higher quality finish is undesirable as the seal may lose drive.

Topic	*Description*	*Recommendations/remarks*
Sleeve sealing and hardfacing	The secondary seal between the shaft and the sleeve (when fitted). 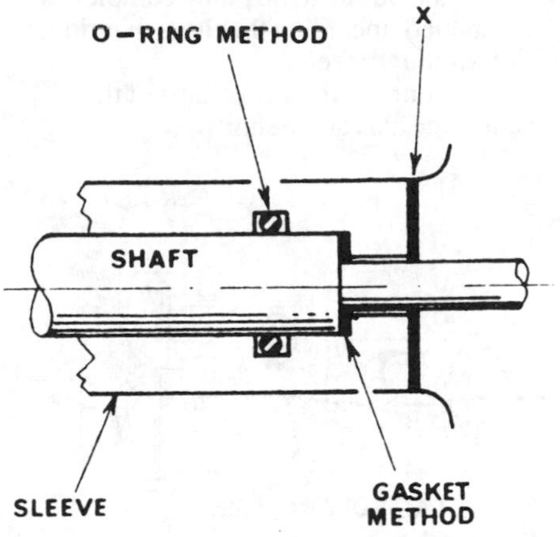	A single 'O' ring *or* a single gasket between the sleeve and shaft step (hook sleeve) is effective provided the sleeve is positively driven (key or grub screws). *Note*: any alteration of gasket thickness can affect seal compression.
	Hardfacing in the secondary seal area. *Note*: this should not extend to grub screw area as drive grip may be impaired. 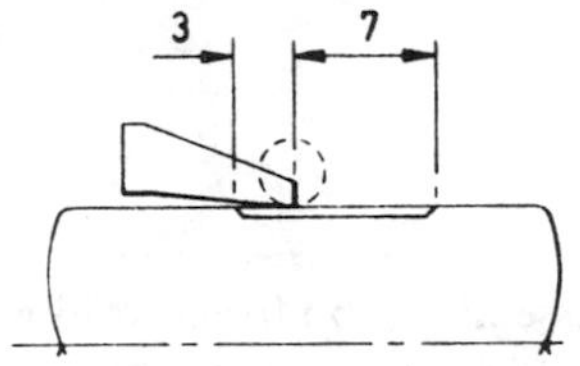	For seals with dynamic 'O' rings or PTFE wedges, the position where this operates on the shaft must be free of scratches, indentations, and other damage which can cause fretting in operation. It is therefore usually recommended to hard face to a minimum of 500 Brinell for a minimum length as shown in the diagram, using a ceramic coating with a ground surface finish as recommended by the seal supplier. Minimum porosity is important for longest lives.
Sleeve fit	Radial fit of sleeve to shaft.	A tight sliding fit with relief along much of its length to ease installation/removal. Typically, H7/g6 fit (*BS 4500, Part 1*) is used up to 4500 r/min.
Concentricity of seal chamber bore	Concentricity of the diameter used to locate the stationary seat (either directly or in a carrier). If another diameter is used, e.g., an outer diameter of the seal chamber, this should be checked. 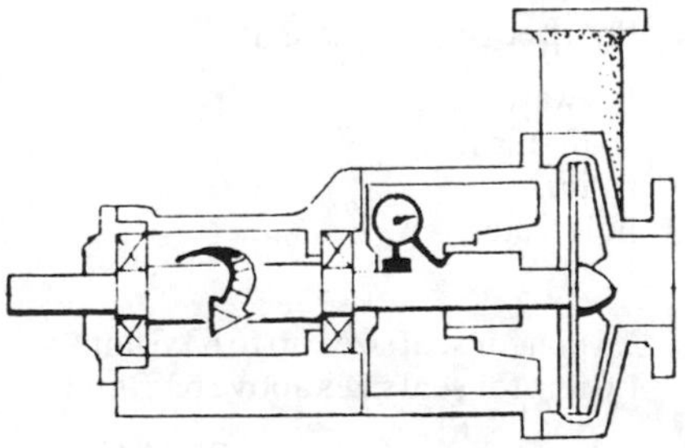	Not to exceed 0.08 mm/m up to a maximum of 0.1 mm (0.004 in) TIR for shaft speed ⩽1800 r/min. Not to exceed 0.04 mm/m up to a maximum of 0.05 mm (0.002 in) TIR for shaft speed >1800 r/min. This allows for the clearance between the shaft and a stationary seat. On double ended pumps the bearing bracket can, if necessary, be moved and redowelled to centralise the shaft in the stuffing box. On pumps previously sealed by gland packing, some cleaning of the box prior to taking the reading may be necessary because of corrosion, etc.

Topic	Description	Recommendations/remarks
Squareness of the stuffing box	The squareness of the face of the seal chamber to the axis of the shaft. Carried out with the pump completely assembled, including the thrust bearing, but without the seals. This controls the squareness of the seat to the axis of the shaft. SEAT SQUARENESS	Typically not to exceed 0.08 mm/m up to a maximum of 0.1 mm (0.004 in) TIR for shaft speed ≤1800 r/min. Not to exceed 0.04 mm/m up to maximum of 0.05 mm (0.002 in) TIR for shaft speed >1800 r/min. See graph for further details. If a seat carrier is used for push-in (inserted and floating) seats, any out of parallel between the locating face of the seat carrier and the back of the seat recess must be added to that of the seal chamber. The total should be within the above limit.
Overall shaft concentricity	On final assembly at both seal position and shaft drive end.	0.08 mm (0.003 in) TIR.
Sharp edges	No sharp edges are acceptable over which a seal must pass with an interference fit. Sharp edges can occur at: keyways; shaft steps; splines; holes, etc. 2.5 mm for seal sizes up to 63.5 mm 4.0 mm for seal sizes above 63.5 mm 10° BREAK EDGE	Remove sharp edges where possible. A typical lead over a shoulder is a 10 degree chamfer of at least 3 mm. Simple guides can be made to protect a critical seal over a particularly difficult shaft. Another technique is to temporarily fill using, for example, fast cure epoxy resin, and then file to shape thus effectively removing a risky obstruction.

9.3 SEAL INSTALLATION

This section is not a step by step guide to fitting every type of seal. It cannot be. It simply lists some useful hints and tips.

Topic	Description	Recommendations/remarks
Part number	A double check that the seal is correct for the duty.	A safety issue in many cases. The numbers should agree on the arrangement drawing, the seal packaging and the seal itself.
Fitting instructions	As supplied with the seal.	To be read and followed.
Arrangement drawing	Gives both dimensional and technical details required.	Some vendors supply one with every seal, others on request.
Seal dimensions checking	A check that dimensions of both seal and seat are correct to the drawing.	It is worth checking that the seal is able to compress to the correct length. Some metal bellows seals mechanically yield if completely compressed during assembly. If this happens the seal will not return to its original length undamaged. Therefore, if in doubt, it is preferable to only compress such seals to the working length.
Seal cavity dimensions	A check that the dimensions of the space into which the seal and seat fit are correct to the drawing.	
Compression length tolerance	To avoid over compression, seal positioning needs to take into account circumstances where this can easily occur, for example: long vertical pumps; multistage between bearing horizontal pumps.	Typically ±0.5 mm (0.020 in). Some seal designs offer extended compression tolerances. Cartridge designs with a facility for setting the seal axially offer a useful solution.
Seal reference dimensions 	A reference mark scribed on the sleeve or shaft so that the position of the seal can be calculated from the drawing.	A mark in line with the face of the seal chamber is a useful reference. When equipment is being dismantled, it is advisable to make this mark wherever possible. Where appropriate use dimples, keys, pins, etc. to ensure correct location.
Stepped shaft for balanced seals 	Many balanced seals require a step in the shaft because the running face is smaller than the diameter on which the seal fits.	The steps in the carbon and the shaft can be quite close and incorrect shaft positioning can damage or even break the carbon. *Note* especially on double ended pumps where excessive shaft movement must be prevented.

Topic	Description	Recommendations/remarks
Metal bellows seal axial resultant force	Although pressures balance each side of each bellows lamination, there is a resultant force as there is one extra side in one direction. The direction depends on the detailed design.	When using, say, a pin drive rather than conventional grub screws, ensure there is a method of fixing the seal axially on the shaft/sleeve otherwise the seal may move axially on the sleeve in service.
Clamped seats	Spigot location of clamped seats. 	Spigot at one location only where the seal maker has specified a dimensional tolerance. Typically such seats are a loose fit in the clamp plate and spigotted into the seal chamber. In this layout clamp plate bolt holes also require an adequate clearance.
	Clamp plate bolt tightening.	It is important not to overtighten clamp plate bolts otherwise seal distortion and hence leakage can result. Typically, sequential tightening is a good practice, e.g., 1, 3, 4, 2 on a four bolt fixing.
Push-in seats	Seat location.	These seats are located with an interference fit on the secondary seal rather than the seat itself. If the seat is a force fit, distortion and leakage can occur.
	Seat anti-rotation pins. 	Anti-rotation pins are worthwhile in the vast majority of cases. It is important to line up the pin and corresponding hole in the back of the seat, before pressing in the seat. If the pin does not enter the hole or if the pin bottoms in the hole or slot, the seat will be lifted out of square and hence give a short seal life. This can be quite difficult to both avoid and notice in a gland plate with a deep seat recess.
'O' rings and similar secondary seals	Renewal.	Replace them each time a seal is dismantled.

	Fitting.
	
		Avoid kinks/twists when in position, e.g., do not roll an 'O' ring on to a stationary seat, but gently stretch the loop over. Always be alert to possible cuts or nicks when fitting past holes, shoulders, keyways, etc.
		    Preform PTFE 'O' rings in boiling water if necessary.
		    The shaft or sleeve should be lubricated, but care must be taken to avoid incompatibility with the elastomer; e.g., mineral oil grease must not be used with EP rubber. Silicone grease is satisfactory for most materials, but must not be used with silicone rubbers. If there is doubt, use water or a soft soap. Avoid getting grease on the seal faces.

Topic	Description	Recommendations/remarks
Auxiliary gland pipework	Pipework for the supply/return, for example, of: product recirculation; jacket cooling water; water, steam, or other flush/quench connections; water, steam, or other for cooled or heated seats; external sealant; drains; vents, etc.	Ensure pipework and connections are clear. Ensure pipework is correctly positioned. Ensure pipework is correctly connected as per arrangement drawing, e.g., steam in at top/water in at bottom. Check sealant supply system is correctly commissioned (critical instrument checks carried out, valves open, etc.).
	New installations.	Check drain connection is at lowest point in the seal plate. Check that a strainer, e.g., a witches' hat, is in the pump suction, to avoid large objects entering the pump. Check that a strainer is fitted into the seal recirculation line to remove initial solids, such as pipescale and dirt. This is best removed when the system is clean. Do not leave strainer in too long as this may block the circulation flow. A full commissioning trial of separate sealant systems (testing relief valves, etc.) is usually worthwhile.
Shaft coupling alignment	Driver to pump shaft alignment during installation. Some pump users consider this so important that periodic alignment checkups are made.	Extremely important as coupling misalignment directly affects seal life. Alignment must be within manufacturers' specifications after making all pipework connections and removing any isolation spades. Misalignment puts lateral, angular and axial loads on the shaft, giving additional bearing wear and failures, and additional shaft deflection/movement at the seal. This swash and track run out at the seal faces reduces life. Even when flexible shaft couplings are well within their capacity, misalignment gives loads, e.g., radial loads from gear couplings.

Topic	Description	Recommendations/remarks
	Loading from main pipework. connections	Misaligned pipework can distort the pump and reduce seal life. Avoid levering, e.g., with a scaffold pole, to line up the suction and discharge connections. The usual technique is to carry out the coupling alignment check before connecting the pipework, leave the dial indicators in position and reconnect the pipework. If there is any variation then it is necessary to cut and reset the pipework until the bolting can be freely fitted without an alignment change.
	Loading from auxiliary pipework connections.	Although the same techniques as above can be used, it is worth highlighting separately, as a small pump can be distorted by large auxiliary pipework used to, say, conform to a site standard. Clearly such pipework needs to fit freely, especially where it is seal welded for toxic/flammable duties.
	Hot and cold alignment checks, typically where pumped fluid temperature is 80°C above or below ambient.	Check alignment immediately on stopping after several hours running at working temperature. Realign if necessary. Various proprietary techniques are available for measuring the cold to hot growth to see if there is a concern.

9.4 SEAL REMOVAL

The first consideration is safety. The pump may have handled a toxic substance and hence both adequate cleaning and care are required during disassembly and inspection. Usually seal removal is the reverse of installation.

Topic	Description	Recommendations/remarks
Safety	Retention of toxic or hazardous substances in the seal cavity, associated components and/or barrier fluids of double and tandem seals. Such toxic or hazardous substances now often require written safe handling procedures as a statutory obligation.	Ensure adequate equipment preparation and cleaning for maintenance plus the availability of necessary safety equipment. The essential precautions include training and informing all who will handle the parts, e.g., suppliers.
Removal instructions	Specific manufacturer removal instructions.	Available/essential for some designs. Otherwise reversal of installation is usually adequate.

Topic	Description	Recommendations/remarks
Reconditioning	Possible for many parts, especially cost effective for expensive designs.	A little care during dismantling can save replacing complete components.
Failure evidence	The condition of the seal components is often the best guide to the cause of unacceptable failures (see Chapter 10).	Note the condition of the seal before removal. Store it carefully for later detailed examination. Check for signs of wear, damage or deposits on any parts being reused. Note any unusual operating or other details on Failure Record Sheet (see section 10.1.2). Involve the engineers from the manufacturers.
Seal reuse and inspection	A wear pattern which microscopically mates the two seal faces is established after a period of operation. It is virtually impossible to replace seals with the original wear patterns of the faces in register. Disturbing the seal in any way will most probably require the establishment of a new wear pattern.	Unless absolutely necessary, do not open the seal faces for inspection after a period of operation. If seal faces are to be reused, they need to be relapped to a level of flatness as achieved during original manufacture (see Chapter 10, section 10.6).

9.5 SEAL MODIFICATION

Mechanical seal modifications can be divided into two categories: firstly, major redesigns and, secondly, smaller changes to the seal itself. The former may involve selection of a new type of seal, redesign of the installation, modifications to change the conditions in the stuffing box, etc. As this effectively means that a completely new seal selection is required, it is useful to refer to earlier sections of this document.

In general, in-house modification of seals is to be avoided. There are times when facing a particular problem, however, that the user may wish to do some experimentation. The following notes give some guidance, but it is recommended that any changes proposed should be discussed in detail with the seal manufacturer. *Users are also advised that any unauthorised modification of the seal will affect the contractual position with their suppliers.*

Topic	Description	Recommendations/remarks
Safety	Safety implications of any modification if the pump is handling a toxic or hazardous fluid. The seal is a key part of the mechanical integrity of the pump.	Consult the manufacturers whenever possible. Modifications without his agreement will invalidate any warranties. If appropriate, hazard and operability studies should be carried out. It is important to beware that any changes do not make matters worse through unforeseen effects.
Changes to metal parts	Machining of metallic seal components such as drive rings, collars, pins, springs, etc.	To ensure standardisation of future spares, it is usually preferable to machine the pump components rather than make non-standard seal parts. Standard machining tools, feeds and speeds can be used dependent on the specific metal being machined.

Topic	Description	Recommendations/remarks
Carbon face machining	This may be required to either decrease the face width or increase the balance. To obtain new dimensions of a machined carbon: New outside radius $= R_0$ $= b_r B_r [1 \pm \sqrt{\{1 + (1/B_r) + (B/b_r B_r)\}}]$ New radius $= r$ $= R_0 - b_r$ where B_r = required balance ratio B = existing balance ratio b_r = required face width.	Tool: single point tipped tool (diamond or carbide or ceramic), e.g., Carboloy 833. The cutting edge dressed to a small nose radius with 10 to 20 degree side relief angle and with a clearance angle dependent on the diameter of piece being machined. Feed: 0.02–0.2 mm per pass (0.001–0.01 in per pass) Speed: 2–5 surface m/sec (400–900 surface ft/min) Note that tool wear is quite high (carbon–graphite used in seals is typically equivalent to Rockwell C62, i.e., a relatively hard tool steel).
Hard face modifications	Changes to material used for harder face, typically silicon carbide, tungsten carbide, alumina, stellite, etc.	The only activity commonly carried out by a user is to lap seal faces, with typically a diamond compound, to refinish the face surface. Machining usually requires grinding with a diamond wheel, spark erosion, or similar techniques and seal manufacturers have the most experience to carry out this successfully on the relatively brittle materials.

9.6   COMMISSIONING AND OPERATION

If a mechanical seal is to provide a good life expectancy with minimum leakage then not only must it be properly designed and installed, but it also requires to be correctly commissioned and operated.

The main aim of seal commissioning is to ensure that the seal does not initially run dry. The aim of continuing seal operation is to maintain the particular environment which has been established as correct for the seal.

For ease of reference, this extensive subject is divided into six areas:

Pre-commissioning – pump
Pre-commissioning – motor
Pre-commissioning – seal
Commissioning
Operation
Standby/storage

9.6.1   Pre-commissioning – pump

Topic	Description	Recommendations/remarks
Avoiding dry running	Ensure there is liquid around the seal before starting the equipment. (*Note* that some seal designs have dry running capabilities. Unless confirmed by the seal manufacturer, however, always assume that a liquid is essential at the seal face.)	It is a prime consideration that a mechanical seal is immersed in a liquid from the very beginning so that it will not be scored or damaged from dry operation. If a squealing noise occurs in the seal area, this indicates insufficient liquid at the seal faces. The installation then needs to be checked immediately to determine how more liquid can be brought to or retained in the seal area.

Topic	Description	Recommendations/remarks
New plant build dirt and debris	Dirt accumulates in systems during construction and seldom can this be completely eliminated. A large percentage of seal failures occur at new plant initial start up; often users store a large start-up spares holdings.	Thorough cleaning and washing of lines and equipment prior to start up can avoid many seal failures. On new plant initial start-ups, as well as witches' hat strainers in pump suctions to catch rags, pieces of wood, etc., it is often worth using fine strainers, cyclone separators and even filters on the seal circulation inlet line of critical duties to remove pipescale, dirt, etc. When the line is clean the filter/strainer element should be removed. If not, it is forgotten, and then it may cause a seal failure later.
		Another technique is to install soft gland packing in the pump for start up and change to a mechanical seal later.
		For lines where solids are continuously present in the circulation to the seal, the strainer needs to be regularly cleaned or, more preferably, a cyclone separator used.
Venting the stuffing box	The pump stuffing box always requires to be vented prior to start-up.	Ideally the installation design should allow the seal space to be vented automatically during pump priming.
		Even though the pump has a flooded suction, it is still possible that air may be trapped in the top portion of the stuffing box after the initial liquid purge of the pump.
		It is especially important on vertical installations, where the seal chamber is the uppermost portion of the pressure containing part of the equipment, to avoid a vapour lock.
Pump priming	It is worth reviewing the installation for the need to positively prime it.	Priming may be necessary in applications with a low or negative suction head.

9.6.2   Pre-commissioning – motor

Topic	Description	Recommendations/remarks
Motor rotation checking	Checking motor rotation by electrical craftsmen, etc.	Although certain specialised seals can be run without liquid, in general this should be avoided. All personnel should be cautioned not to run equipment dry when checking motor rotation. A very brief 'jog' of a few seconds will not harm the seal, but full speed for several minutes under dry conditions may well destroy the seal faces.

9.6.3 Pre-commissioning – seal

Topic	Description	Recommendations/remarks
Seal environment controls	Mechanical seals are often installed with various seal environmental control pipework: product recirculation; jacket cooling water; water, steam or other flush/quench connections; water, steam or other for cooled or heated seats; external sealants; drains; vents, etc.	Check that required pipework is correctly installed as per seal arrangement drawing, and that required services are operating (isolating valves open, control/relief valves properly adjusted, instrumentation checked out, etc.). Note especially that external sealant systems (for double and tandem seals) need to be fully checked out before operating the pump. On back-to-back double seals, the sealant pressure must exceed the product pressure opposing the seal before the shaft is rotated. On circulation connections with heads greater than 3.5 bar (50 lb/in^2) a flow controller is usually recommended. Such controls should be used fully open for a few hours after start up and then adjusted to give equal gland plate and pump casing temperatures.
Cooling systems	Such systems are often used to protect the organic materials in the seal and ensure correct seal performance.	Cooling liquid supply should be left on at all times, if possible (to avoid the risk of forgetting to recommission). This is especially true where hot product may be passing through standby equipment while it is not on line. On hot services which are regularly shut down, it is recommended that the cooling system be left on at least long enough to cool the seal below the temperature limits of the organic materials used in the seal.
Warm-up procedures	To ensure that any products that will solidify are fully melted down before start-up.	Thorough warm-up checks are essential to avoid the seal being stuck up by waxy or solidified materials. Usually it is advisable to leave all heating arrangements on during shutdown to ensure a liquid condition of the product at all times. This helps quick start ups or changeovers. However, with difficult applications (e.g., volatile plus waxy product with low suction head), it is necessary to ensure that venting is carried out to remove vapour from the seal chamber.
Chilling procedures	To ensure that the products, e.g., liquid petroleum gases, are in the liquid state at the seal area.	Start up is usually the most critical period and adequate time is essential for chill down. During operation frictional heat from the seal faces adds to the concern. The seal chamber pressure must be sufficient to prevent vaporisation in the vicinity of the seal faces (see Chapter 1, section 1.3.2).

9.6.4 Commissioning

Topic	Description	Recommendations/remarks
Start-up	Not just operating the machine, but also a careful monitor by sight, noise, touch, etc. is required to check for any immediate concerns. Any available instrumentation can also be used where appropriate.	Start a machine in accordance with the manufacturer's instructions. For example, the usual preferred method for a centrifugal pump is with open suction and closed discharge valves, followed by gently opening the discharge valve.
Loss of suction	Dry operation from loss of suction sometimes occurs, especially at start-up. Common causes are inadequate pipework and process changes.	This needs to be corrected immediately to prevent seal face damage or failure. If suction is lost at start up, it is advisable to again vent the seal chamber prior to restarting the pump. If investigation reveals suction pipework inadequacies, this is ideally rectified early as it has a detrimental effect on both the seal and the pump.
Running-in (seal leakage at start-up and after shutdowns)	Allowing a reasonable period of operation for a seal to adjust itself if it leaks slightly at start-up.	A slight leak which gets progressively less is indicative of leakage across the seal faces and continued running may well cure face damage. 　Liquids that contain good lubricating properties will tend to take longer to wear in the seal than those with lesser qualities. 　When a seal starts with a leakage that remains constant with continued running, it suggests damage to the secondary seal or out-of-flat faces. In this case inspection and repair is required (see Chapter 10).
Vibration	Excessive vibration reduces seal life.	Detailed vibration study is a subject in itself. If a problem occurs at start up and the earlier installation recommendations have been followed, then suspect a foundation problem (e.g., holding down bolt tightness or system resonance) or problems originating in the process. After a period in service, out of balance is the most common cause, e.g., from fouling or erosion. Alignment can also change (see below).
Alignment	Shaft coupling alignment can change in service and reduce seal life either through excessive shaft deflection or excessive vibration.	As well as cold to hot changes (see section 9.3), any form of looseness can cause a change. Alignment can change for a range of reasons, from bolts working loose to excessive structural deflections on marine applications. Some causes can be subtle, e.g., foundations moving with the tides on coastal installations.
Thermal expansion	Differential thermal expansion between components.	This can give quite subtle problems, which become more manifest during start up, shutdown or process transients. Distortion at the seal face only needs to be of the order of 1 μm to give a sealing problem.

9.6.5 Operation

The following points may cause seal failure particularly in the early stages.

Topic	Description	Recommendations/remarks
Process	Upsets or transient plant conditions may cause seal problems even when the seal is correctly suited to the usual operating parameters.	The range of possible upsets is both large and dependent on the particular industrial process involved. Whatever the process cause, some typical results are itemised below.
	High seal cavity pressure.	Usually, high pressures are well catered for, but transients, e.g., back surge from rapidly closing valves, are becoming more common with computerised control of plant.
	Low seal cavity pressure.	Although more subtle than over-pressure, it will give problems, e.g., vaporisation at the seal faces.
	Pump operation at or near shut off conditions (closed valve).	In this situation the most common concern is over-heating of product in the seal area. Pump design, product, and pumping temperature all have a significant input to the sensitivity to this problem. Modern single stage end-suction pumps are less prone to shaft deflection and over-heating, when operating at or near closed valve, than older designs.
	Cavitation, typically from NPSH problem.	Cavitation causes seal, as well as pump, damage. Typically this involves shaft vibration, shaft deflection, bearing failures, wear ring pickup and over-heating in the seal area from lack of product.
	Running dry.	This is a very common cause of seal failure; several of the other effects, e.g., cavitation, overheating, etc., can give lack of liquid at the seal faces.
	Excessive temperature.	There are a wide variety of causes of excessive temperature. The possibility of variation in fluid temperature from any transient or off-load operation should be considered among others.
	Low temperatures giving freezing/solidification.	Waxy solids, icing, or similar effects will interfere with correct seal operation. This freezing can occur in the product, a sealant or the local atmosphere (single seals on cryogenic duties are often quench purged with nitrogen to keep them dry).
	System fouling, reaction products, etc.	As with the above example, there are many ways that solids can interfere with seal operation, for example, because of a process upset or from product reaction with the environment.

Topic	Description	Recommendations/remarks
	Overspeed.	Usually this results in a temperature concern in the seal area.
	Instrument fault.	These may cause process upsets, which give seal problems, e.g., levels, pressures, etc., may give seal problems directly (e.g., on sealant systems) or may give a fault concern where one does not exist. Systematic fault finding using multi-skill men or teams is the usual method adopted.
	Incidental shut off of services	Every effort is required to ensure that valves are not closed on sealant, quench, cooling supplies, etc. Handwheel removal is common. On critical duties, interlocks which prevent pump operation when services are inoperative are sometimes used.
	Human error	There is a complete spectrum of possibilities from closing the wrong valve to a fault in some sophisticated computer software.

9.6.6 Standby/storage

Topic	Description	Recommendations/remarks
Standby equipment	Operating procedures for installed spare pumps fitted with mechanical seals.	To ensure availability of the spare, it is usually wise to operate the spare pump in some way. This not only checks the seal, but also the bearings, coupling, motor, etc. The recommended practice (where practicable) is to use the main and spare pumps alternately at some predetermined frequency. If this is not possible, it is worth just rotating the pump for a few moments, say, once a week. (The frequency may well be set by other constraints).
Long term pump storage	When a seal is installed in pump which will not be used for a prolonged period, for example, on mechanical inventory build up prior to installing a large project or when 'mothballing' a plant LONG TERM HORIZONTAL STORAGE	Storage up to 2 years. (1) Drain and dry out process liquid and seal areas. (2) Cover/plug all equipment openings. (3) Fill stuffing box with recommended lubricant (e.g., oil, glycerine, ethylene glycol, etc.). Avoid petroleum based fluids if EP (ethylene propylene) rubber seals are fitted. (4) Plug all openings to gland/seal chamber. (5) Mask (e.g., with tape) the clearance between the shaft and seal plate to prevent dirt entry. (6) Turn over the shaft several revolutions every three months.

Topic	Description	Recommendations/remarks
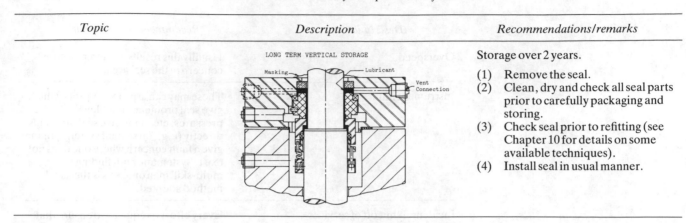		Storage over 2 years. (1) Remove the seal. (2) Clean, dry and check all seal parts prior to carefully packaging and storing. (3) Check seal prior to refitting (see Chapter 10 for details on some available techniques). (4) Install seal in usual manner.

The kind permission of the following Companies to reproduce illustrations used in this Chapter is gratefully acknowledged:

Crane Packing Limited
Durametallic UK
Flexibox Limited

Chapter 10

FAILURE DIAGNOSIS

B. J. Woodley

Analysis of the cause of failure of a mechanical seal often gives the best indication of action required to increase seal life expectancy in a particular situation. The existing seal life may require to be extended for safety or economic reasons.

This section is aimed at an engineer who wishes to carry out diagnosis of his own plant and equipment. It also provides information which will be found useful in discussion with the seal maker to whom troublesome seal failures are usually referred.

The topic is divided as follows:

10.1 Failure definitions, diagnostic approach and recording.

10.2 External symptoms of seal failure.

10.3 Checks before dismantling.

10.4 Checks during dismantling.

10.5 Visual seal examination.

10.6 Other seal examination techniques.

A tabular format is used for ease of reference and section 10.5 has pictures of 45 common failure modes for direct visual comparison.

10.1 FAILURE DEFINITION, DIAGNOSTIC APPROACH, AND RECORDING

10.1.1 Definition

Typically a mechanical face seal is considered to have failed when leakage past the seal assembly causes either:

excessive loss of fluid from the system being sealed;
excessive reduction of pressure within the system being sealed;
excessive addition of barrier fluid into the system being sealed (double seal installations).

In practice, the difficulty is in defining 'excessive'.

Seals functioning perfectly leak to some extent and theoretical definitions of failure are often based on multiples of this value (see Chapter 1, section 1.4.1). In practice, the definition is usually based on plant operator inspection. Typically, a frequent drip is a cause for concern, unacceptable drip frequencies being reduced for toxic or hazardous substances or failures which are known to escalate rapidly. For non-critical duties, e.g., water, much larger leaks may well be accepted before repair. Thus, practical seal failure definition is currently as much dependent on operator psychology and experience as technology.

10.1.2 Diagnostic approach and recording

Seal failure diagnosis is very similar to any other failure investigation and the key message of this section is '*Do not throw any evidence away*'. Often the best indication of the cause of failure is from visual examination of the seal itself and, once the likely cause of the problem is decided, the available solutions are usually clear.

Unfortunately, if evidence is lost in disassembly, there is no way to back track. To reduce the risk of losing some key information, it is therefore suggested to review the failure as follows:

External symptoms of seal failure	(Section 10.2)
Checks before dismantling	(Section 10.3)
Checks during dismantling	(Section 10.4)
Visual seal examination	(Section 10.5)
seal faces	
secondary seals	
seal hardware	
Other seal examination techniques	(Section 10.6)

The mechanical seal failure record sheet shown in Fig. 10.1 is intended to assist in recording the details of the failure and reduce the risk of losing any important information.

MECHANICAL SEAL FAILURE RECORD SHEET   Sheet 1 of 2

Please complete all boxes
(with sketch if appropriate)

/	= Yes
x	= No
–	= No Info

Sheet 1 for General Facts

* = Delete not applicable

Sheet 2 for Visual Examination
        Details

COMPANY	SITE	DATE
PLANT	SERVICE	PLANT REF NO:
MACHINE	SEAL MANF	SEAL TYPE

REASON FOR SEAL REMOVAL   TOXIC/HAZARDOUS PRODUCT?   YES/NO *

EXTERNAL SYMPTOMS OF SEAL FAILURE   SEAL LIFE   HOURS * DAYS STARTS

BACKGROUND INFORMATION

FLUID SEALED	MACHINE DRG NO:
FLUID PRESSURE AT SEAL	SEAL DRG NO:
FLUID TEMPERATURE AT SEAL	BOILING POINT OF FLUID AT SEAL CHAMBER PRESSURE
FLUID FLOW RATE IN SEAL CHAMBER	SHAFT SPEED
	MACHINE VIBRATION

SPECIAL OPERATING CONDITIONS (PROCESS CHANGES, ETC)

SEAL LEAKAGE PATTERN   HYDROSTATIC TEST RESULT

POSSIBLE LEAK PATHS
(Please continue completion
of this box as dismantling
proceeds).

DIMENSIONAL CHECKS

SEAL WORKING LENGTH	SHAFT END PLAY
SQUARENESS OF SEAL FACES TO SHAFT AXIS	SHAFT RUN-OUT & DEFLECTION
CONCENTRICITY OF SEAL FACES TO SHAFT AXIS	OTHER FITTING INFORMATION

DEPOSITS & DEBRIS   SEAL HANG-UP?

VISIBLE DAMAGE TO SEAL FACES   YES/NO *

SEAL PARTS RETURNED TO MANUFACTURER   YES/NO *

GENERAL COMMENTS
(Please continue
on a separate
sheet if necessary)

Fig. 10.1. Mechanical seal failure record sheet

MECHANICAL SEAL FAILURE RECORD SHEET Sheet 2 of 2

VISUAL EXAMINATION DETAILS

DATE:
PLANT REF NO:

PRIMARY SEAL FACES	STATIONARY FACE	ROTATING FACE
MATERIAL		
FACE FLOATING?	YES/NO *	YES/NO *
CONTACT PATTERN		
FRACTURE, SCRATCHES, CHIPS, ETC.		
WEAR, GROOVING AND EROSION		
WEAR DOWN (Please tick box and comment on evenness, etc)	< 0.1 mm / 0.1 - 0.5 mm / 0.5 - 2.0 mm / > 2.0 mm	
THERMAL DISTRESS		
CHEMICAL ATTACK		
OTHER		

SECONDARY SEALS	STATIONARY FACE SECONDARY SEAL	ROTATING FACE SECONDARY SEAL	SLEEVE SEAL	GLAND PLATE SEAL
TYPE & LOCATION (Please note according to design)				
OMITTED/ MISASSEMBLED SEALS				
PHYSICAL DAMAGE				
THERMAL DISTRESS (EG HARD/CRACKED/ SET ELASTOMER)				
CHEMICAL ATTACK				
OTHER				

SEAL HARDWARE	SLEEVE	SPRING(S)	ROTATING BODY
TYPE & LOCATION (Please note according to design)			
OMITTED/ MISASSEMBLED PART(S)			
PHYSICAL DAMAGE			
THERMAL DISTRESS			
CHEMICAL ATTACK			
OTHER			

SPECIAL CHECKS

EG   PRESSURE TEST
PARAFFIN TEST
OPTICAL TEST

10.2   EXTERNAL SYMPTOMS OF SEAL FAILURE

A useful indication of the cause of a seal problem can often be obtained by analysis of the symptoms experienced in service. These may suggest either the remedy directly or at least the direction of subsequent failure diagnosis. On critical duties, instrumentation may be available to give further assistance or portable devices can be used for condition checking.

Symptom	Possible causes	Recommendations/remarks
Seal squeals during operation	Inadequate amount of liquid to lubricate seal faces. (Note that not all dry seals squeal.)	If not in use, a bypass flush line may be required. If already in use, the line or associated restrictions, e.g., orifices in the gland plate, may need to be enlarged.
Carbon dust accumulating on outside of seal area	Inadequate amount of liquid to lubricate seal faces.	See above.
	Liquid film vaporising/flashing between seal faces. In some cases this leaves a residue which grinds away the carbon–graphite seal ring.	Pressure in seal chamber may be excessively high for the type of seal and the fluid being sealed. See below for actions against vaporisation.
Seal spits and sputters in operation (often called popping)	Product vaporising/flashing across the seal faces.	Remedial action is aimed at providing a positive liquid condition of the product at all times. (1) Increase seal chamber pressure if it is possible to remain in seal operating envelope. (2) Check for proper balance design with seal manufacturer. (3) Change to a seal design not requiring so much product temperature margin (ΔT). (4) If not in use, a bypass flush line will be required. (5) If already in use, the bypass flush line or associated restrictions may need to be enlarged. (6) Increase cooling of seal faces. (7) Check for seal interface cooling with seal manufacturer. *Note* that a review of balance design requires accurate measurement of seal chamber pressure, temperature, and specific gravity of product.
Seal leaks and ices seal plate	Product vaporising/flashing across the seal faces.	For remedial action, see above. *Note* that icing may score seal faces (especially carbon–graphite). They should therefore either be relapped or renewed before starting up after the vaporising condition has been rectified.
Seal drips steadily	If possible, first determine the source of the leakage. Heavy leakage is normally from the faces rather than 'O' rings, etc.	
	Primary seal concerns: Faces not flat. Faces cracked, chipped or blistered. Distortion of seal faces for thermal or mechanical reasons (usually determined from wear pattern on faces).	Typical actions for such concerns are as follows. (1) Check for incorrect installation dimensions. (2) Check for improper seals or materials being used in the application. (3) Check gland gasket for proper compression. (4) Check for gland plate distortion because of over torquing of gland bolts (this can cause faces to become distorted). (5) Clean out any foreign particles between seal faces. Relap faces or renew. (6) Check for any installation or similar damage and renew if necessary. (7) Check for squareness of stuffing box to shaft and similar equipment condition concerns. (See Chapter 9, section 9.2.)

Symptom	Possible causes	Recommendations/remarks
		(8) Ensure pipe strain or machine misalignment is not causing distortion of seal faces (especially end suction overhung type pumps). (9) Improve cooling flush lines.
	Secondary seal concerns: Secondary seals nicked or scratched during installation Leakage of liquid under pump shaft sleeve. Overaged 'O' ring. Compression set of secondary seals (hard and brittle). Chemical attack of secondary seals (soft and sticky).	Typical actions for such concerns are as follows. (1) Renew secondary seals. (2) Check for proper lead in chamfers, burr removal, etc. (3) Check for correct seals with manufacturer. (4) Check for correct seal materials with manufacturer.
	Seal hardware concerns: Spring failure. Erosion damage of hardware. Corrosion of drive mechanisms.	Typical actions for such concerns are as follows. (1) Renew parts. (2) Check for improved material availability. (3) Modify recirculation flow arrangement to reduce high velocity jets on hardware. Install cyclone separator to remove solids from recirculation flow.
Pump/shaft vibration	Misalignment Impeller/shaft system imbalance. Cavitation. Bearing problem.	This will reduce seal life even though leakage may not be immediately apparent. See Chapter 9, sections 9.2, 9.3, and 9.6 for details.
Short seal life	Equipment mechanically out of line (e.g., from undue pipe strain).	See above. In the extreme this can cause rubbing of the seat on the shaft.
	Abrasive product (causing excessive seal face wear).	Typical actions are aimed at determining the source of abrasives and preventing them accumulating at the seal faces. (1) If abrasives are in suspension, bypass flushing over the seal faces will improve the situation by keeping the abrasive particles moving and so reducing their tendency to settle out or accumulate in the seal area. A cyclone separator is often added to this bypass line (filters give longer term problems unless regularly cleared). (2) When abrasives are forming locally in the seal area, a bypass flush will help introduce the maximum product to the seal cavity at the correct temperature. Abrasives form in the seal area because of the process liquid cooling down and crystallising or partly solidifying, or because of local product evaporation.

Symptom	Possible causes	Recommendations/remarks
	Seal running too hot.	(1) Check all cooling lines are connected and operational. (2) Check that flow is not obstructed in cooling lines or jackets (e.g., from scale formation). (3) Increase the capacity of cooling lines. (4) A recirculation or bypass flush line may be necessary. (5) Check for possible rubbing of some seal component on the shaft (see also 'misalignment' above). Some good points to check are: neck bush clearance; clearance between the rotating seal unit and the seal chamber bore; the bore of the seat, and the seal plate clearance from the sleeve.
	Inadequate seal type or seal material for duty.	If there is a concern, advice is readily available from seal manufacturers. Seal materials deficiencies may well result in deterioration from corrosion or excessive heat.

10.3 CHECKS BEFORE DISMANTLING

In addition to noting any seal failure symptoms, other checks prior to disassembly can be valuable, either directly or to facilitate later diagnosis. Most of these checks are straightforward and are carried out as routine by most engineers. Thus they are presented as a check-list to act as an 'aide-memoire'.

Topic	Checklist
Toxic/hazardous product	In such cases, all necessary precautions to be observed prior to and during assembly.
Service life of seal	Hours of operation. Duty cycle, stop/starts, etc.
Process change	Identify any change – often the key to a solution. Seal may have been selected on *theory* of process, not practice. Changes in fluid pressure, temperature, or composition. Process variation or fluctuation.
Background information required	Fluid sealed (including contaminants). Fluid pressure on seal and in system. Fluid temperature at seal and in system. Fluid flow rate within the seal chamber. Sealed fluid vapour pressure/temperature data. Operating shaft speed(s). Special operating conditions. Machine assembly drawing. Seal assembly drawing. Seal design data.
Machine vibration	Useful even when not immediately apparent as a symptom. Axial and radial bearing housing or shaft vibration. Frequency analysis to confirm out-of-balance, misalignment, etc., until machine can be stopped for physical checks.

Topic	Checklist
Seal leakage pattern	*Safety Note:* all necessary precautions must be observed during any leakage checks, especially if the fluid is toxic or hazardous. Amount and nature of abnormal leakage? Leakage constant or variable? Leaks when shaft is stationary? Leaks when shaft is rotating? Related to changes of speed, pressure or temperature of operation?
Possible leakage path(s)	An assembly drawing is of great assistance. If possible, identify source of abnormal leakage while equipment is still operating. Inspect exposed machine surfaces for indications of leakage path(s), for example, along shaft, under sleeve, from seal plate gasket, etc. This inspection to continue through subsequent equipment and seal dismantling until the leakage path(s) are all found. Typical leakage paths: face leakage; secondary seal on sealing ring; secondary seal on seat; seal/gasket on seal plate(s); seal/gasket under shaft sleeve; cracked or damaged housing component.
Hydrostatic testing	If possible, for example with double seals, bench testing of equipment can be a useful method of identifying the leak path. With other seal layouts, a suitable test fixture for subassembly pressure testing may be justifiable if large numbers of seals are being examined.

10.4 CHECKS DURING DISMANTLING

A checklist of points worth noting, divided into three categories; general, premature failure, and mid-life failure checks.

10.4.1 General checks

Topic	Checklist
Seal surfaces	Avoid disturbing the seal surfaces. Avoid wiping or cleaning the faces more than is necessary for safe disassembly. Visual examination of seal faces is included in section 10.5.
Dimensional checks	The necessary marks and measurements to determine seal working length; squareness of seal faces to shaft axis; concentricity of seal faces to shaft axis; shaft end play; shaft radial run out, whip and deflection.
Possible leakage path(s)	Examination of surfaces as they become exposed for all possible causes of abnormal leakage.
Deposits and debris	Examination prior to cleaning for: foreign contaminants; wear debris; small fragments or chips from broken components; corrosion products; miscellaneous debris/deposits.
Seal hang-up	Check for hang-up by flexing the seal slightly above and below its installed working length.
Seal sub-assembly cleaning	Avoid removing or obscuring any vital evidence on the seal failure mechanism (especially on the seal faces). Avoid using wire brushes, sharp tools, abrasive cleaners or powerful solvent cleaning agents (which can attack the elastomeric components).
Packaging	For seal manufacturer examination/repair. Many seal makers will personally collect unusual/critical seals for failure diagnosis. Packaging needs to be of high standard (as for new seals). Avoid wire mounted identification tags, etc., which can damage parts in transit.

10.4.2 Premature failure checks

Topic	Checklist
Seal faces	Examination for nicks, scratches and fractures: low power magnification can assist; see also sections 10.5 and 10.6. Examination of non-uniform contact pattern: dirt trapped between the faces; distortion of one or both faces; improperly finished faces; see also Section 10.6 re optical flat checking. Examination for thermal distress: from running dry; heat checks/thermal cracking; pitting, grooving, galling, spalling, blistering, etc.
Secondary seals	Examination for: omitted seals; misassembled seals; nicks, scratches, cuts, and tears; twisted, extruded, or distorted static seals; score marks from relative rotational movement between secondary seals and mating surface; excessive volume change or compression set; fretting of sealing surfaces at secondary seal positions.
Drive mechanism	Examination for: misassembly; misindexing; omission. Check for loss of secondary seal interference when used for drive purposes, e.g., static seals and bellows.
Face loading hardware	Examination for: incorrect type; misassembly; misindexing; omission.

10.4.3   Mid-life failure checks

Topic	*Checklist*
Seal faces	Examination for: overall corrosion; leaching; abnormal grooving; erosion damage; excessive pitting, galling, and spalling; thermal damage such as waviness, heat checks, cracks, blisters, deposition of solid materials, and overall thermal discolouration. Wear profile check by: naked eye examination; use of low incidence angle light to highlight features; $10\times$ magnification, then $50\times$; measurement to determine the amount of wear.
Secondary seals	Examination for: extrusion; chemical attack on both seal and its interface surfaces; excessive volume damage; excessive compression set; hardening and cracking.
Drive mechanism	Examination for: failure; excessive wear. Check for loss of secondary seal interference when used for drive purposes, e.g., static seals and bellows.

10.5   VISUAL SEAL EXAMINATION

This section has illustrated examples of typical seal concerns to allow direct comparison with a problem being experienced. *OFTEN THE BEST INDICATION OF THE CAUSE OF FAILURE IS FROM VISUAL EXAMINATION OF THE SEAL ITSELF.*

The symptoms experienced may not be the prime cause of failure. It is often necessary to identify the root cause in order to avoid a recurrence. Once the likely cause of the problem is decided, the available solutions are usually clear. There are cases, however, where further checks are necessary to clarify diagnosis. There are also proven remedies for particular concerns. Therefore, this section notes likely causes, further checks, and proven remedies, as appropriate, for each symptom.

As there are a relatively large number of ways a mechanical seal can fail (this section lists forty-five), it is helpful to group them alpha-numerically, as shown in Table 10.1 below. This split is somewhat arbitrary and several failure modes are caused by a complex mixture of mechanical, thermal and/or chemical aspects. However, it does show a pattern which can be helpful when using the subsequent extensive table of common seal failure modes. The latter table is similarly divided into three parts; seal faces, secondary seals, and seal hardware.

Table 10.1.

	Contact pattern	Mechanical	Thermal	Chemical
Seal faces	A1: Proper contact pattern.	A11: Fracture.	A17: Thermal distress, over 360°.	A21: Carbon chemical attack.
	A2: No contact pattern.	A12: Scratches and chips.	A18: Thermal distress over 120°–180°.	A22: Corrosion of metal faces.
	A3: Heavy outside diameter contact.	A13: Adhesive wear.	A19: Thermal distress in patches.	A23: Corrosion of hard faces.
	A4: Heavy inside diameter contact.	A14: Abrasive wear.	A20: Coking.	A24: Flaking and peeling.
	A5: Wide contact pattern.	A15: Grooving and severe wear.		A25: Crystallisation.
	A6: Eccentric contact pattern.	A16: Erosion of carbon ring.		A26: Sludging.
	A7: Contact with one high spot.			A27: Bonding.
	A8: Contact at two or more high spots.			A28: Blistering.
	A9: Contact through 270°.			
	A10: Contact at gland bolt locations.			
Secondary seals	—	B1: Physical damage.	B4: Hard or cracked elastomer.	B6: Elastomer chemical attack.
		B2: Extrusion.	B5: Compression set of elastomer.	B7: Corrosion at secondary seal interfaces.
		B3: Excessive torque.		
Seal hardware	—	C1: Physical damage.	C8: Overheated metal components.	C9: Corrosion of seal hardware.
		C2: Hardware rubbing.		C10: Excessive deposits.
		C3: Erosion or abrasive wear.		
		C4: Drive failure.		
		C5: Spring distortion and breakage.		
		C6: Seal hang-up.		
		C7: Sleeve marking and damage.		

Common seal failure modes – seal faces

Symptom	Characteristics	Example	Causes/Checks/Remedies
A1: Proper contact pattern	Typical contact pattern of a *non-leaking* seal. Full contact through 360 degrees on the seat surface with little or no measurable wear on either seal ring. If leakage is present, suspect the secondary seals, and in this situation the seal typically drips steadily with the shaft stationary or rotating.		*Cause* Any leakage is typically from secondary seals. *Checks* (1) Secondary seals nicked or scratched on installation. If so, renew seals, having checked for proper lead in chamfers, removed burrs, etc. (2) Check secondary seals for damage, porosity, thermal or chemical attack. (3) Check for compression set of 'O' rings. (4) Check for correct materials with seal manufacturers. (5) Seal hang-up (see C6 below). (6) Pipework distortion.
A2: No contact pattern	This indicates that the rotary face is not turning against the stationary face.		*Cause* Possibilities include the following. (1) Improper installation. (2) Slipping of the rotary drive mechanism. (3) Interference of a rotary with a stationary component, e.g., seal body with seal chamber bore. (4) Loss or lack of anti-rotation pin in the seat.
A3: Heavy outside diameter contact (Negative coning or rotation)	Heavy contact on the sealing ring and the seat at the outside diameter of the sealing plane. Fades away to no visible contact at the inside diameter of the contact pattern. Possible edge chipping on the outside diameter of the sealing ring. Leaks steadily at low pressure – little or no leakage at high pressure.		*Cause* Usually caused by the faces not being flat because of over-pressurisation of the seal. *Checks* Can also occur from: (1) incorrect lapping leaving the seal faces not flat; (2) excessive swell of confined secondary seals; (3) improper seal face support surface; (4) entrapment of foreign particles; (5) thermal effects (usually on ID; see A17–A19 below).

A4: Heavy inside diameter contact (Positive coning or rotation)

Heavy contact on the sealing ring and the seat at the inside diameter of the sealing plane. Fades away to no visible contact at the outside diameter of the contact pattern. Possible edge chipping on the inside diameter of the sealing ring.

Seal leaks steadily when the shaft is rotating and usually no leakage when the shaft is stationary.

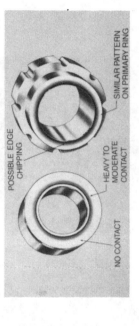

POSSIBLE EDGE CHIPPING

NO CONTACT

HEAVY TO MODERATE CONTACT

SIMILAR PATTERN ON PRIMARY RING

Cause
Typically caused by thermal distortion of seal faces.

Checks
Also can occur from causes listed above under Heavy outside diameter contact, A3.

Remedial actions
(1) Improved cooling of the seal.
(2) Changes of seal material.

A5: Wide contact pattern

Contact pattern is considerably wider on the seat than the face width of the sealing ring. Possible wear at drive notches if present in sealing ring.

Seal does not leak when shaft is stationary, but leaks steadily when shaft is rotating.

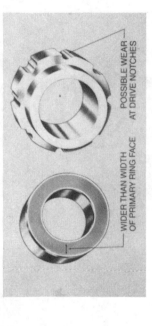

WIDER THAN WIDTH OF PRIMARY RING FACE

POSSIBLE WEAR AT DRIVE NOTCHES

Cause
Possibilities include the following.
(1) Pump misalignment – this may also cause seal to hang-up on the shaft.
(2) Pipe strain.
(3) Bearing failure or excessive clearance.
(4) Bent shaft.
(5) Shaft whirl of large amplitude.
(6) Pump cavitation.
(7) Pump vibration.
(8) Misaligned seat.
(9) Pump operation outside specification.

A6: Eccentric contact pattern

Eccentric contact pattern on the seat with width of contact equal to sealing ring through 360 degrees. Seat may have contact marks on its internal bore or local cracking (from a shaft rub). No abnormal wear on sealing ring if seat is undamaged.

No leakage if the shaft has not contacted the inside diameter of the seal. If seat is damaged, then leakage will occur when the shaft is rotating or stationary.

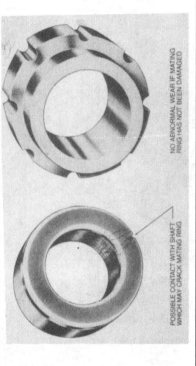

POSSIBLE CONTACT WITH SHAFT WHICH MAY CRACK MATING RING

NO ABNORMAL WEAR IF MATING RING HAS NOT BEEN DAMAGED

Cause
Usually caused by a misaligned seat.

Checks
(1) Check for correct seat design and clearances.
(2) Check for correct clearances between the gland plate and the seal chamber.
(3) Check for concentricity between the outside diameter of the shaft sleeve and the inside of the seal chamber.

Continued

Common seal failure modes – seal faces—*Continued*

Symptom	Characteristics	Example	Causes/Checks/Remedies
A7: Contact with one high spot	Contact pattern on seat through 360 degrees slightly larger than the sealing ring face width. High spot or highly polished area may be present on the seat (for example opposite a drive pin hole or at location of anti-rotation pin if not correctly assembled into hole). Seat without static seal(s) will rock or move in gland plate or holder. Wear at drive notches if present in sealing ring. Seal does not leak when shaft is stationary, but leaks steadily when rotating.	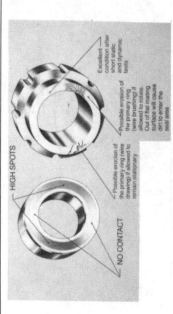HIGHLY POLISHED AREA POSSIBLE OPPOSITE DRIVE PIN HOLE — SLIGHTLY WIDER CONTACT AREA ON MATING RING — WEAR AT DRIVE NOTCHES	*Cause* Mating surfaces are not square. *Checks* (1) Check the seal plate surface in contact with the seat is free from nicks/burrs and shows a full pattern when blued with seat. (2) Check that anti-rotation pin is correctly located into seat. (3) Check that anti-rotation pin does not bottom into the seat. (4) Check for correct extension of all drive pins from seal plate. (5) Check for adequate shaft alignment (to avoid it passing through the seal chamber at an angle). (6) Check for piping strain on pump casing.
A8: Contact at two or more high spots	Seat is distorted mechanically, typically creating two large contact spots – pattern fades away between contact areas. Sealing ring shows excellent condition after short static and dynamic tests. Possible wire drawing erosion of the sealing ring if it remains stationary. Possible wire brushing erosion if the sealing ring rotates, because out-of-flat mating surface allows dirt to enter the seal area. Seal leaks steadily when the shaft is rotating or stationary.	HIGH SPOTS — NO CONTACT — Excellent condition after short static and dynamic tests — Possible erosion of the primary ring if allowed to remain stationary — Possible erosion of the primary ring (wire brushing) if allowed to rotate. Out of flat mating surface will cause dirt to enter the seal area Uneven wearing of face due to distortion	*Cause* Seal faces not flat. *Checks* (1) Check for seal plate distortion because of over-torquing of bolts. (2) Check flatness of faces using optical flat (see section 10.6). (3) Check squareness of parts used to clamp seat. (4) Check seal chamber face flatness of split case pumps. (5) Check the seal plate surface in contact with the seat is free from nicks/burrs and shows a full pattern when blued with the seat.

A9: Contact through 270 degrees

Seal is distorted mechanically giving contact through approximately 270 degrees with the pattern fading away at the low spot.

Sealing ring shows same symptoms as for mechanical distortion above.

Seal leaks steadily when shaft is rotating or stationary.

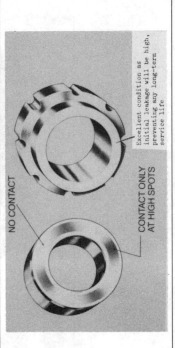

NO CONTACT

CONTACT PATTERN

Possible erosion of the primary ring (wire drawing) if allowed to rotate. Out of flat mating surface will cause dirt to enter the seal area

Possible erosion of the primary ring (wire drawing) allowed to remain stationary under pressure

Excellent condition after short static and dynamic tests

Cause
Seal faces not flat.

Checks
(1) As above for Contact at two or more high spots (A8).
(2) Review the possibility of high seal chamber pressure.

A10: Contact at gland bolt locations

Seat is distorted mechanically giving high spots at each bolt location.

Sealing ring in good condition as initial leakage is high, preventing any long term service life. Seal leaks steadily when the shaft is stationary or rotating.

NO CONTACT

CONTACT ONLY AT HIGH SPOTS

Excellent condition as initial leakage will be high, preventing any long-term service life

Cause
Seal faces not flat.

Check
(1) Check for seal plate distortion because of over-torquing of bolts.

Remedial actions
(1) Change to a softer gasket material between the seal chamber and the seal plate.
(2) Provide full face gasket contact or contact above centreline of bolts to prevent bending of the seal plate.

A11: Fracture

Broken seal rings or cracked seal rings (if retained in some assembly). Many seal face materials are brittle and relatively thin sections are fragile.

Non-uniform discoloration or partial discolouration of the fracture surface or the presence of wear debris indicates fracture prior to or during seal operation. If no wear debris is present, the fracture probably occurred during disassembly. Fractures caused by excessive face torque generally emanate from one or more points of drive engagement and also show wear or damage on mating drive device. This problem can occur when PTFE 'O' rings are used to seal a pinned stationary carbon seat without a buffer sleeve over the pin. In this case, it can result in a severe gouge emanating from the pin slot rather than ring fracture. Seal leaks steadily when the shaft is stationary or rotating. When broken parts are well retained the amount of leakage can sometimes be remarkably low.

Cause
Possibilities include the following.
(1) Mishandling before or during assembly.
(2) Improper seal assembly or installation.
(3) Excessive face torque: jamming from improper assembly; failure of axial holding devices; excessive fluid pressure; poor lubrication; corrosion at seal faces; pin sleeve of PTFE not fitted as recommended by seal makers.
(4) Excessive hydraulic pressure.
(5) Excessive swell of confined secondary seals.
(6) Damage during seal removal and disassembly.
(7) Excessive thermal stress from thermal shock or excessive gradients (see Thermal distress, A17–A19 below).

Continued

Common seal failure modes – seal faces—Continued

Symptom	Characteristics	Example	Causes/Checks/Remedies
A12: Scratches and chips	Scratches in the radial direction usually give a leak regardless of depth or width. In other directions, scratches less than 1 μm deep by 25 μm wide do not typically cause extensive leakage. Scratches and nicks are often erroneously cited as a cause of seal failure and it helps to decide if the scratch was caused before, during or after operation. If the wear pattern is altered by the scratch, then the scratch occurred before or during operation. If the same scratch extends outside the mating area, it is more likely to have occurred prior to operation. If it does not extend outside the mating area and is spiral in form relative to the shaft axis and in the direction of rotation, it probably occurred during operation and can be attributed to a particle entering or coming from the seal faces. Scratches that interrupt, but do not alter, the wear pattern, were probably produced after seal operation. Chips are usually at seal face edges and severe chipping is similar to that caused by excessive hydraulic distortion. Leakage rate depends on the degree of damage and may reduce when the shaft is stationary.		*Cause* Possibilities include the following. (1) Mishandling during manufacture, storage, assembly or installation. (2) Dirt trapped between seal faces. (3) Edge chipping from slamming together during operation when pump cavitates or fluid vaporises at seal faces. *Checks* Edge chipping can also occur from the following. (1) Excessive shaft run out. (2) Excessive shaft deflection or whip. (3) Out of square seal faces. (These conditions also give excessive wear of the drive mechanism.)
A13: Adhesive wear	A combination of mild adhesive/abrasive wear is the normal way seals wear out over a long service life (see Proper contact pattern, A1). Excessive adhesive wear leaves typical non-metallic seal faces heavily worn with a relatively smooth appearance and a minimum of grooving. Severe adhesive wear of metallic faces can lead to scuffing, grooving, and even face seizure. Seal leaks when shaft is rotating. When stationary, the seal may hold or may leak severely.		*Cause* (1) Inadequate lubrication. (2) Excessive seal contact pressure for the face materials. (3) Degraded seal face conditions. *Checks* (1) Check for excessive local temperatures caused by inadequate cooling for the face surface speed. (2) Check *PV* value of seal face materials (this method has its limitations: see Chapter 1, section 1.3.1). *Remedial actions* (1) Improved seal lubricating properties can be achieved by a temperature change. (2) Changing seal face materials. (3) Changing seal balance.

A14: Abrasive wear

Excessive abrasive wear leaves seal faces severely grooved and even scuffed (both metals and non-metals). Harder faces show regular grooving, whilst carbon faces tend to wear less evenly with heavy scoring both across the face and in the direction of rotation. Virtually no wear takes place away from the face contact. Mild abrasive wear from very fine particles gives a wear pattern similar to adhesive wear.

The key clue to abrasive wear is the deposit of solids on the seal faces or adjacent to them. The solids may also result from chemical effects (see A20, A21, A22, A23, A24, A25, A26, A27, and A28 below).

Seal leaks steadily when shaft is stationary or rotating.

Cause
Pumped product contains abrasive matter of a size able to enter between the faces and cause wear.

Remedial actions
(1) Introduce a clean flow to the seal through a cyclone separator (if particle size within separator specification).
(2) Introduce a clean flow to the seal from a separate source.
(3) Install harder wear resisting face materials, e.g., silicon carbide, tungsten carbide.
(4) Use double seal.

ABRASIVES IN PRODUCT RESULT IN GROOVING ACROSS FACE ON CARBON BODY

ABRASIVES IN PRODUCT RESULT IN REGULAR CIRCULAR GROOVING ON A HARD-FACED BODY

A15: Grooving and severe wear

High wear, even cracking, of the seat with polished circumferential scoring, discoloration, and over-heating symptoms. Metal parts may 'blue' with heat of dry running. Even short periods of dry running can form a deep wear groove.

The sealing ring displays severe, though even, wear throughout 360 degree, with gramophone scoring'. Soft carbon sealing rings possibly have edge chipping. Harder sealing rings, e.g., tungsten carbide, have rounded edges. Possible wear at any drive mechanism or notches. Other overheating symptoms may be apparent, e.g., hardening and cracking of 'O' rings.

This is often a start-up problem and the seal drips steadily when the shaft is stationary or rotating.

Note that the scoring damage can be confused with abrasive wear (see A14).

Cause
Dry running because of insufficient or no liquid between the seal faces.

Checks
(1) Check for adequate priming and seal chamber venting.
(2) Check pump suction flows and filters.
(3) Check for blockage/restriction of circulation line.

Remedial actions
(1) If a circulation line does not exist, review the need to install one.
(2) Increase seal circulation flow.
(3) Review operating procedures (see also Chapter 9, section 9.5).

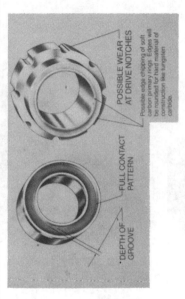

'DEPTH OF GROOVE

FULL CONTACT PATTERN

POSSIBLE WEAR AT DRIVE NOTCHES

Possible edge chipping of soft carbon primary rings. Edges will be rounded for hard material of construction like tungsten carbide.

GROOVING AND SEVERE WEAR

Continued

Common seal failure modes – seal faces—Continued

Symptom	Characteristics	Example	Causes/Checks/Remedies
A16: Erosion of carbon ring	If the carbon ring is on a rotating component, this results in a sculptured appearance with islands of original mating surface still showing. If the carbon is the stationary component, this forms a groove part way across the carbon face adjacent to the circulation inlet on the seal plate. In severe cases harder face materials such as alumina can also be eroded in a similar manner. Seal leaks when the shaft is stationary or rotating.	STATIONARY CARBON ROTATING CARBON	*Cause* Caused by excessive flow velocity at the seal circulation inlet, the circulation flow containing abrasive materials or a combination of these. *Remedial actions* (1) Adding a flow controller in circulation line. (2) Shrouding the seal faces. (3) Injecting the circulation at several points. (4) Methods to reduce abrasive damage as for abrasive wear.
A17: Thermal distress over 360 degrees (vaporisation)	High wear or thermally distressed surface (heat checking) through 360 degrees. This appears as radial surface cracks sometimes accompanied with circular scoring or discoloration from over-heating. If necessary dye penetrant can help to show up the surface cracks. The carbon sealing ring shows high wear and possibly light pitting leading to 'comet' trailing. Possible edge chipping of the sealing ring because of opening and closing of the seal faces and also possible wear of any drive notches. Carbon dust deposits on the atmospheric side of the seal and wear/fretting of the shaft/sleeve at the secondary seal (if dynamic) are also symptoms. Seal leaks steadily when shaft is stationary or rotating. The latter usually with sound from flashing or face popping.	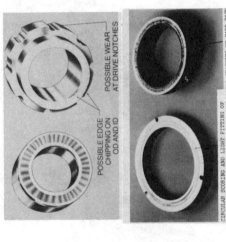POSSIBLE EDGE CHIPPING ON OD AND ID POSSIBLE WEAR AT DRIVE NOTCHES CIRCULAR SCORING AND LIGHT PITTING OF CARBON FACE DUE TO VAPORISATION RADIAL CRACKING ON HARD FACE DUE TO VAPORISATION	

Other seal damage can also result, e.g., fatigue of metal bellows or wear of shaft/sleeve at secondary seals (called 'wedge etching' for PTFE wedge designs). In the latter case, carbon pick-upon to the secondary seal and wear of the secondary seal (e.g., at the nose of the wedge) may be apparent.

Remedial actions
(1) Use a narrow face carbon (of the order of 2.5 mm).
(2) Increase cooling to faces:
   check circulation lines for blockage;
   check cooling coils, lines and jacket for blockage;
   increased circulation flow assists in marginal situation.
(3) Review options to alter seal chamber pressure; on multiple stage pumps the seal chamber pressure may be taken off another stage to prevent flashing. The seal design will require review to ensure it is then not overpressurised.
(4) Review seal design and seal material selection, e.g., use a seal design not requiring so much product temperature margin (ΔT).

A18: Thermal distress over 120 to 180 degrees

Thermally distressed (heat checked) area approximately one-third of the contact pattern. Distressed area 180 degrees from inlet of seal flush with good contact pattern at flush inlet.

High sealing ring wear with possible carbon deposits on the atmospheric side of the seal. Also possible wear at any drive mechanism notches.

Seal drips steadily when shaft is rotating or stationary – possible sound from flashing or face popping.

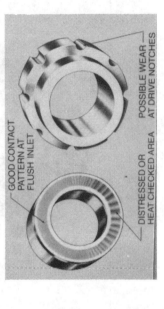

GOOD CONTACT PATTERN AT FLUSH INLET

POSSIBLE WEAR AT DRIVE NOTCHES

DISTRESSED OR HEAT CHECKED AREA

Cause
Sealed liquid vaporising 180 degrees from the seal flush.

Checks
(1) Check for adequate clearances around the seal faces to give sufficient face lubrication and cooling.
(2) Check that seal chamber neck bush clearance is correct.

Remedial actions
(1) Add a circumferential flush groove in the gland plate.
(2) Add a tangential inlet matched to the shaft rotation to aid distribution.
(3) See also under Thermal distress over 360 degrees (A17).

A19: Thermal distress in patches

Two, three, four, five, or six hot spots of thermally distressed or heat checked surface. These patches are sometimes called thermal asperities.

High sealing ring wear with possible carbon deposits on the atmospheric side of the seal. Also possible wear at any drive mechanism or notches.

Seal leaks steadily when the shaft is rotating or stationary. Leakage may be in the form of vapour and with sound from flashing or face popping.

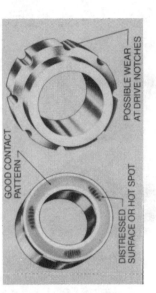

GOOD CONTACT PATTERN

POSSIBLE WEAR AT DRIVE NOTCHES

DISTRESSED SURFACE OR HOT SPOT

Cause
Sealed liquid vaporising between the seal faces. Failure from hot spots is more likely to occur on light specific gravity liquids at high speeds and pressures.

Checks
(1) Check for adequate cooling of seal faces.
(2) Check for seat distortion.

Remedial actions
(1) Increase cooling of seal faces.
(2) Review possibility of seal interface cooling with the seal manufacturer.
(3) See also under Thermal distress over 360 degrees (A17).

Continued

Common seal failure modes – seal faces—Continued

Symptom	Characteristics	Example	Causes/Checks/Remedies
A20: Coking	This usually occurs with hydrocarbon products at high temperatures. It is indicated by failure of the seal to follow up, i.e., no sliding action. This can be found after removal of the seal plate during the stripdown for inspection. Coke particles collect on the inside of the sliding member, even to the extent where it can be difficult to remove. In many cases of continuous operation, heat from the product and seal friction can keep the coke and associated waxes and gums reasonably soft and the seal will operate satisfactorily. Leakage typically occurs on start-up after a period of shut-down or on standby when solidification of waxes/gums associated with the coke particles takes place. The leakage can in odd cases reduce after a short period of running as these waxes soften.		*Cause* Minute quantities of leakage carbonising on the atmospheric side of the seal causing the sliding member to jam and hence not follow up any face wear. *Remedial actions* (1) The usual approach with hydrocarbons is to fit a permanent low pressure steam quench on the atmospheric side of the seal to prevent the build up and solidification of coke and wax particles. An adequately sized drain will both prevent excessive steam pressure and assist particle removal. This quench must be operational before start-up. (2) If not already fitted, a high temperature lip seal at the back of the seal plate improves quenching efficiency. It also reduces the likelihood of steam entering the bearing housing.
A21: Carbon chemical attack	Area of carbon ring in contact with the product is corrosively attacked resulting in overall material removal, pitting, porosity, softening, or disintegration. Essentially, there are two carbon–graphite corrosion modes: overall corrosion and selective leaching. Overall corrosion occurs when it is attacked by highly oxidising acids or highly concentrated caustic fluids. A hardness reduction of 20 Shore scleroscope points is typical for carbon–graphite materials which have been chemically attacked. In severe cases of this type, seal faces are reduced to sludge. Selective leaching of the impregnant (added to the otherwise porous carbon to make it impervious) results in either increased wear rate or seal face porosity. With this mechanism, a hardness reduction of 5 Shore scleroscope points is typical for carbon–graphite materials. Pressure testing for porosity can also be used to confirm such a problem. Seal leaks steadily when shaft is stationary or rotating.		*Cause* Incompatibility of the carbon with the product resulting in two failure mechanisms. (1) Overall corrosion. (2) Selective leaching of impregnant. *Remedial actions* A change of material – both failure mechanisms require checking the material selection for product compatibility and the original product conditions against the seal selection. A corrosion rate of 0.025 mm (0.001 in) per year is normally quite unacceptable for seals, even though this is satisfactory for most industrial hardware. It is usually, therefore, better to use seal manufacturer data than any non-numerical industrial corrosion data when assessing such a problem. Many highly corrosive products, e.g., oleum, present a conflict between corrosion and wear resistance of the face materials which, even with the latest materials, results in a maximum seal life of only a few months.

In some cases, corrosive attack can be disguised when corrosion products act as fixed or free contaminants and abrasively wear one or both seal faces. The corrosion products can also cause seal hang-up (see C6 below).

Cause
Many corrosion failure mechanisms such as overall corrosion, intergranular corrosion, stress-corrosion cracking, etc., occur in mechanical seals.

Remedial actions
This can be analysed and solved in just the same way as with other mechanical devices.

Cause
In typical commercial alumina (75 or 85 percent), the alumina particles are bonded together by a predominantly silica glass binder. Sealed fluids with a pH greater than 10, or containing hydrofluoric acid, leach out this binder giving the failure characteristics described.

Certain grades of silicon carbide contain free silicon which can be similarly attacked (e.g., by hydrofluoric acid). See Chapter 2.

Acidic fluids may leach nickel or cobalt binders incorporated in cemented tungsten carbide, again giving the failure characteristics described.

Remedial actions
99.5 per cent alumina, silicon-free silicon carbide (sometimes called Sintered alpha), and alloy-bonded tungsten carbide are now available that withstand such fluids more effectively.

A22:	Corrosion of metal faces	Corrosive attack by the product, sealant, or atmosphere. Corrosion is accelerated because the face is subject to sliding contact wear. Dissimilar materials can also set up an electrolytic corrosive action. Seal leaks steadily when shaft is stationary or rotating.

A23:	Corrosion of hard faces	This is commonly the result of leaching of binders or fillers in alumina, tungsten carbide, and silicon carbide. Certain corrosive fluids leach the binders/fillers from these ceramics and, in effect, convert the seal face into a grinding surface. As leaching continues the ceramic particles eventually become dislodged from the base material and cause abrasive wear of one or both seal faces. Seal face flatness is degraded to the point of seal failure by the resulting voids in the ceramic surface and/or the abrasive damage. Seal leaks steadily when shaft is stationary or rotating.

Continued

Common seal failure modes – seal faces—*Continued*

Symptom	Characteristics	Example	Causes/Checks/Remedies
A24: Flaking and peeling (of hard coatings)	Stainless steel seal faces are usually plated with a hard facing of stellite, ceramic, tungsten carbide, or a variety of other materials. The failure often starts with slight blistering, then lifting of the coating. Final failure may well be accelerated by abrasive wear of one or both seal faces by hard particles as they become dislodged from the coating. Seal leakage can escalate quickly and continues when the shaft is stopped.		*Cause* Possibilities are: (1) a defective coating; (2) chemical attack at the bond between the base metal and the coating. *Checks* The chemical attack may be aggravated by both heat generation at the seal face and the porosity inherent in some coating techniques. *Remedial actions* Changing to a solid face material is the usual solution adopted.
A25: Crystalisa-tion	Similar symptoms as for Coking (A20), except that it occurs on various products and conditions. Sometimes the crystals embed in the softer face and rapidly abrade the harder face. Note that as well as from the product, crystals can come from the atmosphere (e.g., ice crystals) or from a barrier fluid (e.g., hard water deposit). Leakage rates vary widely.		*Cause* A build-up of crystals from the pumped product giving both high Abrasive wear rates (see A14) or failure to follow up (see Coking, A20, and Seal hang-up, C6 below). *Remedial actions* As with Coking, the best remedy is a permanent quench to dissolve or disperse the crystals. Typical examples of quench fluids are hot water, steam, and solvent, according to the product. Again, lip seal improves quench efficiency. The crystals can come from the atmosphere, e.g., ice on LPG pump duties, where a nitrogen quench to keep moisture from the seal is one possible approach.

A26: Sludging

A polished wear track or slight scoring on the hard face. Small cavity holes on the carbon face (from which particles have been pulled). Possible distortion of the drive spring or excessive wear/damage on other drive mechanisms.

Associated with the sealing of high viscosity liquids, particularly acute on pumps sealing hydrocarbon liquids at temperatures above ambient. When shut down, the viscosity of the pumped liquid and the interface film increases as the temperature drops and problems may arise on restarting the pump.

Once leakage occurs after start-up, it seldom stops when the pump is stopped again.

Cause
The shear stresses between the seal faces exceed the rupture strength of the carbon and particles are pulled from the carbon face. This is usually because of a viscosity increase when shut down, but it also occurs when the interface film partially carbonises from overheating.

Checks
(1) Ensure viscosity range of products is within seal capabilities.
(2) Check that pump head is adequate to give product circulation around the seal area under pumping conditions.

Remedial actions
(1) To overcome start-up problems:
   preheat circulation lines (e.g., by steam tracing);
   preheat seal area (e.g., low pressure steam to seal chamber jacket/tracing);
   preheat seal faces (e.g., low pressure steam quench).
   Such heating to be used for 15–30 minutes prior to start-up.
(2) Supply continuous heat through a heated seal plate.

A27: Bonding

Similar phenomenon to Sludging (A26). In this situation, a bond is formed between the two seal faces after the pump has been stationary for a long period. On starting, particles are pulled from the the carbon face and leakage occurs.

The appearance of the seal and other symptoms are similar to that from sludging problems.

Once leakage occurs after start-up, it seldom stops when the pump is stopped again.

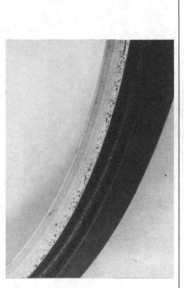

Cause
The main cause is when a pump is tested on a different liquid to that on which it will operate and a chemical reaction occurs between the test fluid film and the actual product film.

Remedial actions
(1) Selection of suitable test fluid.
(2) Operation on an intermediate flushing fluid for a short period between testing and production use.

Continued

Common seal failure modes – seal faces—Continued

Symptom	Characteristics	Example	Causes/Checks/Remedies
A28: Blistering	Similar phenomenon to Sludging (A26) and Bonding (A27). Initially, this failure appears as a shiny bruised effect in the surface, and at a later stage manifests itself as a crater where the bruise has detached itself from the surface and passed through the seal faces. Normally associated with start-stop applications. Seal leaks steadily when shaft is rotating or stationary.		*Cause* High local heating occurs in a few seconds on start-up, particularly with high viscosity products in high speed, motor-driven pumps operating at high pressure. This heating can cause rapid expansion of liquid that has been absorbed into the seal face surface. This rapid expansion causes high stress which, in extreme cases, exceeds the rupture strength of material. *Remedial actions* Difficult problem to solve; useful approaches include the following. (1) Keeping product viscosity low by heating. (2) Careful choice of seal materials. Ones with higher thermal conductivity produce less blistering against carbon counterfaces. Certain grades of carbon–graphite are more resistant. (3) Review of start-up procedures.

Common seal failure modes – secondary seals

Symptom	Characteristics	Example	Causes/Checks/Remedies
B1: Physical damage	Cuts, scratches, nicks, or tears in 'O' rings, bellows, wedges, and other secondary seals. Plastic seals, e.g., PTFE, possess less elastic self-healing properties than elastometric secondary seals. All forms of bellows, rubber, PTFE, and metal, can easily be damaged and the location may not be easy to spot. Seal drips steadily when shaft is stationary or rotating.		*Cause* Possibilities include the following. (1) Mishandling. (2) Inadequate installation practice. (3) Presence of dirt. (4) Failure to remove burrs, sharp edges of steps, keyways, holes, etc., and previous set screw indentations prior to seal installation. (See Chapter 9, section 9.2.) (5) Bellows damage can also be caused by manufacturing defects – inclusions, incorrect curing, inadequate weld quality, etc. *Remedial actions* Having found the cause, the only usual rectification of the secondary seal damage is renewal.

B2: Extrusion

This can occur with 'O' rings, wedges, bellows, and other secondary seals. The commonest form is 'O' ring extrusion, and this occurs when part of the 'O' ring is forced through close clearance gaps. Typically, a lip is first formed on the 'O' ring; it is then cut and in some cases peeled off like an outer cover.

Flaying or shredding is more common on synthetic rubber rings, whereas a lip is usually formed on Viton or PTFE. Thermoplastic materials, e.g., PTFE and Viton, are more susceptible to extrusion at elevated temperatures. Seal leakage may reduce when shaft is stopped.

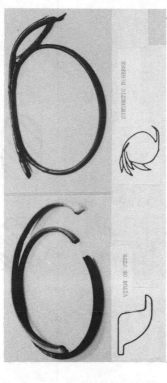

Cause
Possibilities include the following.
(1) Use of excessive force when fitting and assembling components.
(2) Excessive pressure (possibly aggravated by overheating and chemical incompatibility).
(3) Incorrect shaft and/or 'O' ring groove sizing giving excessive clearance between components.

Remedial actions
As well as checking the above, other changes can be made, such as fitting a back-up ring, a change of seal design, a change of material, etc.

B3: Excessive torque

Some secondary seals provide a drive function; exceeding the torque capacity will cause problems. Typically this will either involve (a) rotational movement resulting in wear or ultimate failure of seal from frictional heat developed during sliding contact, or (b) exceeding the structural torque capacity of the device. An example of (a) is rotation of a seat dependent on friction of its 'O' ring to avoid rotation (no anti-rotation pin). An example of (b) is bellows torsional failure. This can give very large seal leaks.

The photograph shows a metal bellows failure (rubber bellows tear in a similar manner). This can be compared with bellows over-pressurisation which can also rupture the bellows.

Cause
Possibilities include the following.
(1) Bonding of a high viscosity film between the seal faces (A27). On start-up the bond strength is greater than the design torque capacity of the seal.
(2) High seal face friction, e.g., from lack of lubrication.

Remedial actions
If the cause cannot be rectified (using methods referred to under Sludging (A26), Bonding (A27), Adhesive wear (A13), Grooving and severe wear (A15), and Thermal distress (A17, A18, A19)); modified seals with an anti-rotation device appropriate to the problem are available.

Continued

Common seal failure modes – secondary seals—*Continued*

Symptom	Characteristics	Example	Causes/Checks/Remedies
B4: Hard or cracked elastomer	Rubber 'O' ring hardened and cracked. PTFE 'O' ring discoloured blue/black. The portion of the ring nearest the faces is usually the worst. Most commonly a problem with nitrile rubber. Comparative analysis of secondary seals from all locations will reveal whether the thermal condition was local to one secondary seal or an overall excessive temperature. It is important to distinguish between chemical attack and thermal damage to decide on the remedy. Chemical attack is more likely on secondary seal surfaces exposed to the fluid; thermal degradation is more frequently found on surfaces exposed to the atmosphere. Seal drips steadily when shaft is stationary or rotating.	 OVERHEATING OF 'O' RING RESULTS IN CRACKING AND PITTING OF SURFACE	*Cause* Two possibilities; overheating or chemical attack; see also Elastomer chemical attack (B6 below). If most or all damage is on secondary seal surfaces that contact a seal face member, excessive frictional heat from the faces is the likely cause. Other possible thermal damage sources are: heat soak from the seal environment including the shaft and housing; relative rotational movement between the secondary seal and the shaft or housing. It is important to identify the source of thermal damage as it may lead to the root cause of the failure. For example, excessive loading of the seal face material could have caused the frictional heat, and changing the 'O' ring material would not avoid premature future failure of the seal faces. *Checks* (1) Check circulation to seal area. (2) Check for dry running, low pump suction flow, sludging, etc. (3) Ensure any cooling is fully operational. (4) Check product conditions are as originally specified and that 'O' ring material is suitable.
B5: Compression set of elastomer	Although this will occur over a period of time, early changes in section as shown will result in premature failure. Compression set does not involve a significant volume change. Seal leaks steadily when shaft is stationary or rotating.	 REGULAR COMPRESSION SET	*Cause* Excessive temperature for the 'O' ring material. Sometimes caused by incompatibility with fluids.

Common seal failure modes – secondary seals—Continued

Symptom	Characteristics	Example	Causes/Checks/Remedies
B6: Elastomer chemical attack	This gives excessive volume change, either swell or shrinkage, which causes a seal failure through one or more of the following. (1) Extrusion caused by swell. (2) Seal face distortion and misalignment caused by swell. (3) Loss of secondary seal interference caused by shrinkage. (4) Shrinkage of seals giving loss of secondary seal drive. Leakage may also occur from the 'O' ring being eaten away. It may also appear to have lost its original composition and to be breaking up. Often product side is badly attacked whilst non-product side has a relatively good appearance. Leakage rates vary widely.	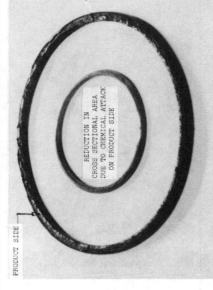 PRODUCT SIDE REDUCTION IN CROSS SECTIONAL AREA DUE TO CHEMICAL ATTACK ON PRODUCT SIDE	*Cause* Chemical attack of elastomer by the product. *Checks* It is necessary to check the original product conditions against the seal selection and ensure that the 'O' ring fitted is made of the correct material. 'O' ring/chemical incompatability charts are available from seal manufacturers. If there is a doubt about a volume change, secondary seal dimensions should be measured in both free and assembled conditions and compared with those specified on assembly drawing. An optical comparator is one useful instrument for such 'O' ring examination. Specially coloured 'O' rings to assist identification help to ensure that the correct material is used. However some colouring additives may have a lower corrosion resistance than the base elastomer.
B7: Corrosion at secondary seal interfaces	This gives a subtle leakage path resulting from two different mechanisms; fretting corrosion and crevice corrosion. Fretting corrosion is caused by small relative movements between a secondary seal and its mating surface. The degree of damage is accelerated in the presence of even a slightly aggressive product (e.g., water) and is particularly aggravated by the presence of chlorides. The fretting corrosion debris is abrasive and the later stages of attack are assisted by a three body abrasive wear mechanism, where debris embeds in the secondary seal and wears the shaft or sleeve. Crevice (or oxygen concentration cell) corrosion occurs because secondary seals, e.g., elastomeric bellows, can trap a small amount of fluid adjacent to the shaft. A good indication of this failure mode is a polished or gas-scrubbed area adjacent to the corroded section which is generated by hydrogen emanating from the crevice, i.e., from beneath the bellows.		*Cause* Fretting corrosion is primarily governed by mechanical factors such as equipment condition, seal assembly procedures and correct materials selection. *Checks* Common contributors to fretting corrosion include the following. (1) Excessive shaft end play – over 0.1 mm (0.004 in). (2) Excessive shaft deflection – over 0.08 mm (0.003 in). (3) Excessive out-of-squareness of seal face to shaft axis – over 0.08 mm (0.003 in) TIR. Fretting corrosion is most common at the dynamic secondary seal under a pusher type seal (for which the above values refer – see Chapter 6, sections 6.2 and 6.3, and Chapter 9, section 9.2, for further details). In a pusher type seal, the secondary seal is pushed along the shaft or sleeve to compensate for wear.

Continued

Common seal failure modes – secondary seals—*Continued*

Symptom	Characteristics	Example	Causes/Checks/Remedies
			Remedial actions (1) It is common to hardface the sleeve in the secondary seal area to minimise the damage from fretting corrosion. (2) Use a non-pusher seal, e.g., a metal bellows seal to avoid the fretting contact. (3) If crevice corrosion is suspected, then any action to avoid the crevice or provide a corrosion resistant surface treatment will correct this effect.

Common seal failure modes – seal hardware

Symptom	Characteristics	Example	Causes/Checks/Remedies
C1: Physical damage	A wide variety of symptoms from chips, minor distortion, nicks in metal bellows, to the example in the picture. In that specific case, care was taken not to damage the faces by placing the seal on its edge. Unfortunately, it was not wedged, it rolled away and was run over by a forklift truck.		*Cause* Not observing good fitting practice: insufficient cleanliness; excessive force; use of incorrect tools, etc.). After being careful with the seal faces and secondary seals, the hardware is sometimes damaged by accident.
C2: Hardware rubbing	Certain conditions may cause abnormal wear where little should occur, e.g.: the outer skin of the rotary unit; the shaft (e.g., against the stationary seat); the neck bush; the throttle bush in the back of the seal plate. In severe cases, the part may be heated to such an extent that it reaches its melting point.		*Cause* Possibilities include the following. (1) Bearing failure. (2) Pump/motor shaft misalignment. (3) Seal chamber too small for rotary unit. (4) Unspigotted stationary unit slips and touches shaft. (5) Non-piloted seal plate touching shaft. (6) Set screws in the rotary unit coming loose and contacting the seal chamber wall. (7) Pieces of the face break off and jam between the rotary unit and the seal chamber wall. (8) Flush connection lines extend too far into the seal chamber and touch the seal.

(9) Single spring seals may rub the seal chamber wall if broken or over-compressed or are subject to high speed.

(10) Multiple springs break up and jam between the rotary unit and the seal chamber wall.

(11) Product or other seal deposits (see C10 below) may scale up on the seal or on the seal chamber wall.

(12) Thermal expansion causing the metal body or other part to expand and hence contact the seal chamber wall.

(13) Equipment vibration.

C3: Erosion or abrasive wear

Circular marks on the outside diameter of the rotating seal body – often in line with a circulation inlet.

On stationary seal hardware, grooving damage occurs again, often in line with a circulation inlet.

Cause

This can be caused by Hardware rubbing (see C2).

It also can result from the incoming flush containing abrasives and eroding the seal body especially if the flush pressure differential is too high.

Also caused by wear debris circulating in the seal chamber

Remedial actions

Solutions can involve:

(1) Changing the circulation inlet position.

(2) Making it tangential.

(3) Checking this inlet for protrusion into the seal chamber.

(4) Flushing with a cleaner fluid.

(5) Selecting a smaller outside diameter seal.

(6) Boring out the seal chamber.

ROTATING HARDWARE

STATIONARY HARDWARE

Continued

Common seal failure modes – seal hardware—*Continued*

Symptom	Characteristics	Example	Causes/Checks/Remedies
C4: Drive failure	This can occur with both the torsional drive devices of rotating components and the anti-rotation devices of stationary components. Typical examples include the following. (1) Wear/fracture of drive pins. (2) Wear of drive lugs. (3) Fatigue failure of metal bellows (an adequate product temperature margin, ΔT, is vital as this is often caused by vaporisation). (4) Failure of drive screws/collars, e.g., set screws cutting into the body.		*Cause* Possibilities include the following. (1) Jammed seal assembly. (2) Excessive shaft end play. (3) Failure of axial holding device. (4) Poor seal face lubrication. (5) Excessive seal fluid pressure. (6) Seal face out of square with shaft axis. (7) Excessive shaft run out. (8) Excessive shaft deflection. (9) Equipment vibration. (10) Stick–slip face friction giving seal face vibration.

C5: Spring distortion and breakage	All mechanical seals require movement to keep the faces together during changing pump and seal conditions and to compensate for wear. Spring action is obtained by a single coil spring, multiple coil springs, a metal bellows assembly, or a wave spring washer.
	    Typical failure characteristics are radial cracking of the spring section (especially on the inside diameter), straight fracture, wear marks in ends of spring coils and on the sleeve and rotary necks, and build up of solid contaminants around spring(s) making them ineffective.
	    See also Excessive torque (B3), *re* bellows assembly failure.
			Cause
			These spring devices fail in a variety of ways, e.g., corrosion, stress-corrosion and fatigue.
			Checks
			On many single spring seals, the drive is unidirectional and the spring should always grip its mating parts. With such seals, reverse rotation or incorrect spring fitting (see Chapter 9, section 9.1) causes the spring to tend to uncoil, slip, distort, crack, or even break.
			    The above and other spring problems are most common on high viscosity duties prone to Sludging (A26) or Bonding (A27).
			    On multi-spring seals, a build-up of solids around the springs can make some springs ineffective and, hence, cause overload and failure of the others.
			Remedial actions
			On multi-spring seals, diversion of part of the product circulation through the spring pockets can reduce future build-up of solids.

FATIGUE   CORROSION   STRESS

C6: Seal hang-up

This occurs when the sliding assembly is prevented from following up (by moving axially), thus leaving a gap between the sealing ring and the seat.

The sliding assembly movement is typically prevented by a build-up of deposited dissolved solids, corrosion, oxidation, or decomposition products. This possibility is present whenever a pusher type seal (where the secondary seal is pushed along the shaft or sleeve to compensate for wear) is used. See also Coking (A20) and Crystallisation (A25).

Remedial actions
(1) Use of a non-pusher type seal, e.g., metal bellows type.
(2) Provision of a suitable quench, for example:
water to prevent deposition of aqueous dissolved solids;
nitrogen to prevent the formation of oxidation products;
oil or similar to prevent the formation of corrosion products;
a suitable coolant quench to prevent thermal decomposition products from forming.
(3) Use of a seal design in which the secondary seals advance on to clean surfaces can also help.
(4) In many cases, mechanical sleeve damage that has occurred will require rectification (including hard facing in the secondary seal area).

GAP BETWEEN FACES OF PRIMARY SEAL RING AND MATING RING

SECONDARY SEAL

STATIONARY MATING RING

SHAFT

PRIMARY SEAL RING

Bold dashed lines indicate solid deposits on shaft

C7: Sleeve marking and damage

This may well relate to Seal hang-up (C6), Coking (A20), or Crystallisation (A25). The marking on a sleeve (or shaft if no sleeve is fitted) often gives a useful indication of the cause of seal failure.

This marking can be divided into three types.

(1) From mechanical reasons – causes in this section give details.
(2) From fretting corrosion or crevice corrosion between the sleeve and the secondary seal – see Corrosion at secondary seal interfaces (B7).
(3) Overall corrosion, usually found on the product side of the sleeve; unless the seal is leaking badly, the atmospheric side is often in good condition. (See Corrosion of seal hardware (C9).)

Mechanical causes typically give leakage only when running and often leakage disappears when the machine is static.

When in operation an increase in shaft eccentricity will increase hydrodynamic action, resulting in thicker fluid film and increased leakage.

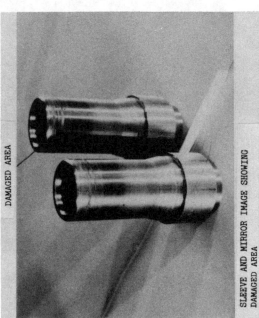

DAMAGED AREA

SLEEVE AND MIRROR IMAGE SHOWING DAMAGED AREA

Cause
Typical causes of sleeve marking.

(1) Contact between 'O' ring landings on the inside of a rotary seal ring is often caused by an eccentric or misaligned shaft. If landing wear is severe 'O' ring extrusion can result.
(2) If contact in (1) above occupies all the sleeve circumference, it is probably caused by a misaligned seat which forces the sealing ring to oscillate relative to the shaft sleeve once per revolution. This is often accompanied by wear on the inside diameter of the secondary seal on the sealing ring.
(3) If the contact in (1) above occupies part of the sleeve circumference, this usually suggests an eccentric or gyrating shaft. It is often an indication that external forces are imposing misalignment between the seal faces and causing leakage.
(4) A shaft which is bent often gives rise to two marks diametrically opposite: one on the front landing and one on the opposite rear landing. This bending may be from an out-of-balance shaft/rotor assembly.

Continued

Common seal failure modes – seal hardware—*Continued*

Symptom	Characteristics	Example	Causes/Checks/Remedies
			(5) Severe vibration (from the above or other reasons) which causes the 'O' ring landings to contact, resulting in fretting and marking into which foreign matter lodges, thus giving Seal hang-up (C6).
			(6) Failed bearings can result in either increased vibration or misalignment.
			(7) Incorrect sleeve manufacture or seal assembly.
			(8) Lack of hard facing giving excessive wear in secondary seal area of shaft, especially if abrasives are present in the product.
C8: Overheated metal components	When steel is heated, a colour change takes place. This heating causes tempering and, hence, loss of required mechanical properties. This colour change may be present generally or related to specific components. Typical colours and temperatures that create these colours on stainless steel. (1) Straw yellow: 370–430°C (700–800°F) (2) Brown: 480–540°C (900–1000°F) (3) Blue: 590°C (1100°F) (4) Black: 650°C (1200°F)		*Cause* An easily distinguishable sign of seal trouble. Unless caused by Hardware rubbing (C2), there are usually other components, i.e., seal faces or secondary seals, which are also damaged and assist diagnosis of the likely cause of excessive heat. Typical reasons are dry running, vaporisation, excessive heat soak, etc. *Checks* Comparative analysis of parts from all locations will reveal whether the thermal condition was local to one component or an overall excessive temperature.

C9: Corrosion of seal hardware

Corrosive attack results in overall and local loss of metal. Damage characteristics are usually indicative of the corrosion mechanism; these mechanisms are as conventionally experienced in other engineering components.

Corrosion damage is often not present on the atmospheric side of the seal (unless it was leaking badly). Seals can continue to function adequately until quite advanced stages of corrosion.

Cause
This occurs through the various usual mechanisms; overall corrosion, stress-corrosion, electrolytic attack, hydrogen embrittlement, crevice corrosion and fretting corrosion.

Checks
(1) Check material selection against the product and its conditions.
(2) Check for correct processing in manufacture with seal supplier.
(3) Review use of any dissimilar metals (electrolytic action).

Remedial actions
(1) A change of material.
(2) Ground the pump effectively to earth if pitting from electrolysis is suspected.

C10: Excessive deposits

Deposits from the product, corrosion, etc., build up on the rotary body. This can cause the rotating unit to 'freeze' in the seal chamber. Other concerns, e.g., Seal hang-up (C6), may well occur first.

Cause
Inadequate seal chamber circulation to flush out the deposits and stop them building up.

Remedial actions
In severe cases, a separate clean purge may be required.

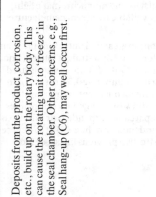

10.6   OTHER SEAL EXAMINATION TECHNIQUES

Although visual examination with the naked eye is the most valuable and widely used method, other methods are available to assist failure diagnosis. The more sophisticated ones can be carried out by the seal manufacturer or other consulting service where it is not worthwhile to do it in-house.

Method	Description	Results interpretation/Comments
Optical magnification and assistance	(1) Conventional magnifying glass. (2) Pocket-sized measurement microscope (up to 50×). (3) Binocular low power microscope (10× – 50×). (4) Dye penetrant for crack detection.	To assist the naked eye. High power magnification techniques, such as scanning electron microscopes, are useful in difficult cases, but much experience is required in interpreting results.
Pressure test	Hydrostatic testing of seal assemblies – relatively easy when process pressure acts on inside of seal as with externally mounted bellows seals.	This often requires a test fixture which may be justifiable if a large number of seals are to be checked. Many users return seals to suppliers for such checks.
Paraffin test	A rotating metal bellows assembly or similar is 'ringed' on to a flat surface and filled internally with paraffin, white spirit or similar. The penetrating power of such liquids acts as a good leak detector.	A remarkably powerful check when conventional vacuum test equipment, as used by the seal maker, is not available. Leaks can be often readily spotted directly in the workshop. If the mechanical seal contains ethylene–propylene elastomeric seals, avoid using a hydrocarbon for testing. Soapy water is an alternative, but is less effective.
Optical flat	This allows accurate measurement of seal face flatness. It is a transparent quartz or glass disc having at least one surface accurately flat. For seal checking, an accuracy range of $0.1\ \mu m$ is adequate, although more expensive flatter ones are available. The diameter of the flat needs to be at least equal to the outer diameter of the seal face being checked. The flat is placed on the surface being checked and dark bands are produced from interference of the light waves reflected at the two surfaces. These dark bands are used to determine the degree of flatness. When interference bands are straight, parallel and equally spaced, the surface is assumed to be flat to within $0.3\ \mu m$ ($11.6\ \mu$ in). With this level of accuracy, clearly both the flat and the seal faces must be absolutely dry and free from dirt, dust, lint, etc. In frequent use, it is convenient to use a monochromatic light source to produce the bands and avoid coloured fringes. Both optical flats and monochromatic light sources are available from seal manufacturers, etc. Interpretation is carried out noting the number of bands intersected by a straight tangent line as in the examples. Out-of-flatness is measured by multiplying this number by $0.3\ \mu m$ ($11.6\ \mu$ in.). If the bands are inconsistent or missing, it is necessary to draw two imaginary centre lines 90 degrees apart and perpendicular to the axis of the part and then draw line AB at 45 degrees connecting the two previous lines (see later examples).	For correct seal face performance, three light bands is typically the maximum flatness tolerance. Flatness between one and two light bands is consistently achieved by reputable seal makers. 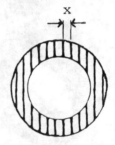 Flat within one light band. The distance x is dependent on the amount of air between the optical flat and the face and does not indicate lack of flatness. 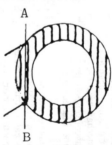 Bands bending at outer edges indicate wash out of the periphery due to the polishing process. As line AB intersects one black band, the face is out by 1 light band, i.e., $0.3\ \mu m$.

Method	*Description*	*Results interpretation/Comments*

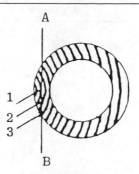

Bands bend on one side and line AB intersects 3 bands. The face is therefore out of flat by 3 light bands or 0.9 μm.

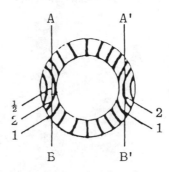

This indicates an egg-shaped curvature of 2½ light bands, i.e., 7 μm. Line AB intersects two bands and falls between another two at the centre of the ring.

   Line A′B′ intersects two bands that curve in the opposite direction.

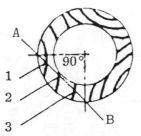

Bands show a saddle-shape out-of-flat condition of 3 light bands, 0.9 μm.

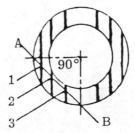

Bands show a cylindrical-shaped part with a 3 light band reading error.

Table 10.6—*Continued*

Method	Description	Results interpretation/Comments
		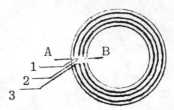 Band symmetrical pattern indicates convex or concave part. The out-of-flatness is measured by the number of bands on the part, i.e., 3 bands or 0.9 μm.

The kind permission of the following companies to reproduce illustrations used in this chapter is gratefully acknowledged: Feodor Burgmann Dichtungswerke GMBH; Crane Packing Limited; Durametallic UK; Flexibox Ltd.

APPENDICES

APPENDICES

APPENDIX 1

GLOSSARY OF MECHANICAL SEAL TERMS

Notes.   The definitions refer solely to the sense in which the term is used in mechanical seal technology.

Terms in italics in the definitions are referred to separately in the alphabetical list.

Terms involving more than one word are listed in the order in which they appear in the text, viz *Operating length* not *Length, operating.*

Terms regularly encountered in connection with mechanical seals, but which have been avoided in this publication are given in parentheses.

Abeyance seal — A non-contacting *Auxiliary seal* that is activated by failure of the *Primary seal* in the case of a *Single seal*, or the outer seal in the case of a *Double seal.*

Abrasive wear — Wear occurring by the mechanical action of an abrasive. Abrasives are substances that are harder than the abraded surface and usually have an angular profile.

Adhesive wear — Wear arising from small-scale local welding at asperities; a common wear mode associated with running in and mild steady state wear.

Anti-rotation device — A device, usually a pin, designed to prevent the *Stationary seal member* from rotating in its mounting.

API 610 — American Petroleum Institute Standard: *Centrifugal pumps for general refinery services* (7th Edn. in preparation). A specification widely used for heavy duty centrifugal pumps.

API piping plan — Arrangements recommended in API 610 for connecting auxiliary pipework to the *Seal chamber.*

Asperity — Minute high spot on the *Seal face* resulting from the manufacturing process.

(*Autobalancing* — Alternative term for *Double balancing.*)

Auxiliary seal — A seal fitted to the atmospheric side of a *Quench chamber* or *Secondary-containment* chamber.

Back-to-back seal — A *Seal configuration* consisting of a *Double seal* with the *Seal rings* adjacent to each other, i.e., two *Mechanical seals* facing in opposite directions).

(*Back-up seal* — Alternative name for *Auxiliary seal.*)

Balance diameter — The smaller diameter at which the *Hydraulic load* operates on an *Internally-mounted seal*; the larger diameter on which the *Hydraulic load* acts on an *Externally-mounted Seal.*

Balance ratio — The proportion of the *Seal chamber* pressure that is applied to the *faces* of a *Mechanical seal.*

Balanced seal — A *Mechanical seal* design in which the *Balance ratio* is less than 1 (typically 0.6–0.7).

Barrier fluid — A liquid injected between a pair of *Double seals.*

Bellows seal — A seal in which one of the *Faces* is mounted on a bellows.

Blistering — A term used to describe a particular form of damage of carbon-graphite *Seal faces*, usually caused by hydrocarbons.

Boundary lubrication — Condition of lubrication where the *Seal faces* are in solid contact though separated by adsorbed surface films.

Bubble point — Mixtures of liquids do not have a clearly defined boiling point. The *Bubble point* is the temperature at which the first bubble is evolved on raising the temperature at constant pressure. The term is most frequently used with mixtures of hydrocarbons.

Cartridge seal — A self-contained *Mechanical seal* that can be bench assembled and fitted directly without the need for further adjustment.

(*Clamp plate* — An alternative term for *Seal plate.*)

Closing force — Combined *Hydraulic* and *Spring load* acting on the *Floating seal member* in the closing direction.

Coking — The formation of carbonaceous deposits on the atmospheric side of a *Mechanical seal* resulting from the oxidation/polymerisation of *Leakage* of organic products.

Compression set — The difference between the thickness of a gasket or elastomer, or length of a spring, as supplied and after being subject to compression in service. More specifically the *Compression set* of an elastomer is defined as:

$$\frac{\text{change in specimen length}}{\text{applied strain}} \times \text{original specimen length}$$

following a specified test procedure (*BS 903*).

Coning — Axisymmetric distortion of the *Seal faces* giving a radial variation in seal *Film thickness*.

(*Contact pattern* — An alternative term for *Wear track*.)

(*Controlled leakage seal* — Alternative term for *Hydrostatic seal*.)

Coolant — A liquid from an external source circulated through a hollow *Stationary seal member* or other separate cooling element to remove heat.

Cyclone separator — A hydrocyclone fitted in a *Product recirculation* line to remove solids.

Crystallisation — The formation of crystalline solids on the atmospheric side of a *Mechanical seal* resulting from evaporation of *Leakage* of *Product*.

Dead-ended — A *Seal arrangement* in which there is no *Product recirculation* or injection of *Flush* into the *Seal chamber*.

Degree of balance — The proportion of the face area that is exposed to the low pressure side of the *Balance diameter* (= 1 − *Balance ratio*).

Delta T, ΔT — The difference between the bulk temperature of the liquid in the *Seal chamber* and the boiling point (or *Bubble point* in the case of mixtures) of this liquid at the pressure in the *Seal chamber*. Also known as the *Product temperature margin*.

Diameter ratio — The ratio (>1) between the outer and inner diameters of the narrower of the *Seal faces*.

Double balancing — A *Mechanical seal* design feature which changes the balance diameter to improve the seal's resistance to operating under reverse pressure. This prevents opening of the inside seal in a *Double seal* upon loss of *Barrier fluid* pressure. (It is sometimes called *Autobalancing*.)

Double seal — Restricted in this publication to the arrangement of two *Mechanical seals* in a *Seal chamber* sealing in opposite directions. The seals can be in either the *Back-to-back*

or *Face-to-face seal* configurations (*qv*).

Note: an alternative usage is to include two seals sealing in the same direction in the category of *Double seal*; in this publication this latter configuration is referred to as a *Tandem seal*.

Drain connection — A connection to the *Quench* (or *Secondary containment*) *chamber* for the collection of liquid.

Drive pin — A device for transmitting torque from the shaft to the *Rotating seal member*.

Dry running — Running with no liquid between the *Seal faces*.

Duty parameter — A non-dimensional parameter used to reflect the lubrication conditions at the *Seal face* contact. It is defined as $G = \eta Vb/F_t$.

Dynamic secondary seal — A *Secondary seal* in a *Pusher seal* that prevents leakage between the shaft or housing and the *Floating seal member* of a *Mechanical seal*.

Early-life failures — Failures occurring shortly after start-up because of manufacturing or fitting errors; sometimes referred to as infantile mortality.

Erosion — *Abrasive wear* of a surface by small particles in a gas, vapour or liquid, or droplets of liquid in a gas or vapour (wire-drawing) flowing across it.

Externally-mounted seal — An arrangement in which the *Mechanical seal* is mounted outside the pump or sealed vessel so that fewer seal parts are exposed to contact with a corrosive *Sealed fluid*. In this arrangement the *Sealed fluid* is in contact with the inner diameter of the *Seal faces*.

Face — Strictly the surface of a *Seal ring* at the *Sealing interface*, but also commonly used for the whole ring, e.g., *Hard face*.

Face load — The combined spring and *Hydraulic load* carried between the *Seal faces* before allowing for any fluid pressure in the *Sealing interface*.

Face-to-face seal — A *Seal configuration* consisting of a *Double seal* with the *Seats* adjacent to each other, i.e., two *Mechanical seals* facing in opposite directions.

Face width	Half the difference between the outer and inner diameters of the narrower of the *Seal faces*.
Film thickness	The thickness of the *Fluid film* between the *Seal faces*.
Film transfer	A process by which a film of the material of the *Soft face* is deposited on the *Hard face*.
(Flashing	An alternative term for *Popping*).
Flatness	The degree of flatness (peak-to-valley amplitude) of the *Seal faces*, normally expressed in helium *Light bands* (1 helium light band = 0.29 μm.
Floating seal member	The spring-loaded seal member of a *Mechanical seal* which is allowed limited axial movement to accommodate shaft end float and seal wear.
Fluid film	A film of liquid separating the *Seal faces*, generated by *Hydrostatic* and/or *Hydrodynamic lubrication*.
Fluid-film lubrication	Condition of lubrication in which the *Seal faces* are completely separated by a liquid film.
Flush	A separate liquid injected to the *Seal chamber* to prevent access of solids or corrosive liquids to the *Seal faces* or to provide cooling.
Flush connection	Connection to the *Seal chamber* to allow circulation of the *Sealed fluid*.
Free length	The unconstrained axial length of a *Mechanical seal*.
Friction coefficient	Defined in a *Mechanical seal* as the ratio of the friction force at the *Sealing interface* to the *Closing force*.
(Gland plate	An alternative term for *Seal plate*.)
Hang-up	Failure of the *Secondary dynamic seal* to move under the applied spring and hydraulic forces.
Hard face	*Seal face* manufactured from ceramic, cermet or metal.
Header tank	An external vessel providing a *Barrier fluid* to a *Double seal*, either with a static head or with a thermal syphon system.
Heat checking	The formation of fine radial cracks on a *Hard face* caused by thermal stresses set up by alternate *Dry running* and quenching by cold liquid as the film is re-established.
(Hydraulic balance	A synonym for *Balance ratio*.)
Hydraulic load	The load on the *Floating seal member* resulting from differential

	pressure between the *Seal chamber* and the low pressure side of the seal acting on the area of the *Sealing ring* above the *Balance diameter* plus that caused by pressure on the low pressure side acting on the area of the *Seal ring* below the *Balance diameter*.
Hydrodynamic lubrication	Fluid-film lubrication in which the pressure in the *Fluid film* is generated by the relative velocity of the *Seal faces*; this can be in either the circumferential or axial direction.
Hydrodynamic seal	A *Mechanical seal* designed to operate with *Hydrodynamic lubrication* between the *Seal faces*.
Hydrostatic instability	*Face* separation occurring when hydraulic opening forces exceed the *Total closing force*.
Hydrostatic lubrication	*Fluid-film lubrication* in which the pressure in the *Fluid film* is generated externally to the *Seal faces*.
Hydrostatic opening force	The separating force on the *Seal faces* resulting from the hydrostatic pressure between the *Faces*.
Hydrostatic seal	A *Mechanical seal* designed to operate with *Hydrostatic lubrication* between the *Seal faces*.
Icing	Build up of ice on the outside of a *Mechanical seal* caused by solidification of atmospheric water vapour through evaporative cooling of *Leakage* of a liquid sealed above its atmospheric boiling point.
Internally-mounted seal	The normal arrangement with the *Mechanical seal* mounted inside the pump or sealed vessel. In this arrangement the sealed liquid is in contact with the outer diameter of the *Seal faces*.
L_{10} *life*	A statistic used to express the life of a population of *Mechanical seals*; it is the time when 10 per cent of the seals have failed.
Leakage	*Sealed fluid* loss from the system; it includes non-obvious vapour formed by evaporation as well as the more obvious liquid emission. Leakage may occur through *Secondary* as well as *Primary seals*.
Light band	Refers to the wavelength of helium light (= 0.29 μm) used as a measure of the *Flatness* of the *Seal faces*.

Mechanical seal	A device for sealing a rotating shaft whereby the *Sealing interface* is located between a pair of radial faces, one rotating, the other stationary.
Mixed lubrication	Condition of lubrication where the load between the *Seal faces* is partly carried by *Boundary lubrication* and partly by *Fluid-film lubrication*.
MTBF	Mean time between failures. A statistic used to express the life of a population of *Mechanical seals*. It is given mathematically by the following expression:

$$MTBF = \frac{L_1 + L_2 + \cdots + L_n}{n}$$

where L_1, L_2, etc., are the lives of individual seals.

Neck bush	Close clearance bush at inner end of *Seal chamber* to restrict flow of dirty fluid from pump into the *Seal chamber* or maintain pressure of *Recirculation flow* in *Seal chamber*.
Net closing force	The difference between the *Total closing force* and the *Hydrostatic opening force*.
Non-pusher seal	*Bellows seal* in which the *Dynamic secondary seal* is eliminated.
Operating length	Axial length of installed *Mechanical seal*.
Optical flat	Glass plate used for measuring *Flatness* of the *Seal faces*.
'O'-ring	Toroidal sealing ring used as *Secondary seal* in both static and dynamic situations.
Popping	A term used to indicate intermittent leakage of vapour characterised by popping sound.
Primary seal	The seal formed by the *Stationary* and *Rotating seal members*. In *Double seals* refers to the seal on the *Product* side.
Product	The process fluid.
Product recirculation	Circulation of the *Product* through the *Seal chamber* to provide cooling. (See *Recirculation flow, Reverse circulation*.)
Product temperature margin	Alternative name for *Delta* T, ΔT.
Pumping ring	A device fitted inside the *Seal chamber* to circulate the liquid in the *Seal chamber* through an external cooler and/or *Header tank*.

Pusher seal	*Mechanical seal* in which there is a *Dynamic secondary seal*, as distinct from a *Bellows seal*.
PV *factor*	A parameter used to express the severity of *Seal face* operating conditions. In this publication it is defined as the product of the pressure drop across the seal and the mean relative velocity of the *Seal faces*.
Quench	Fluid introduced to atmospheric side of *Mechanical seal* to remove toxic or flammable *Leakage* to a safe place or to prevent *Coking, Crystallisation* or *Icing*.
Quench chamber	Enclosed space on the atmospheric side of a *Mechanical seal* to which the *Quench* is introduced; normally fitted with an *Auxiliary seal* to prevent excessive leakage to atmosphere.
Random failures	Failures occurring during operation other than *Early-life failures* and those caused by normal wear-out of the *Seal faces*.
Recirculation flow	Flow of the *Product* from the pump discharge through the *Seal chamber* to the back of the pump impeller or from the back of the pump impeller through the *Seal chamber* to the pump suction.
Reverse balancing	Selection of the *Balance diameter* so that a *Mechanical seal* can withstand pressure on the inside diameter of its face rather than the outside diameter, i.e., the reverse of normal outside diameter pressurisation. This is of particular use for the inboard seal of a *Double seal* as it puts any solids on the outside diameter of the inboard seal and minimizes clogging.
Reverse circulation	Flow of the *Product* from the back of the pump impeller through the *Seal chamber* to the pump suction to provide cooling of the seal and reduce access of solids to the *Seal faces*.
Rotating seal	*Mechanical seal* in which the *Floating seal member* is mounted on the shaft.
Rotating seal member	The seal member that is mounted on the shaft.
(Rotation	Alternative term to *Coning*.)
Rotor	*Rotating member* of *Mechanical seal*.

Seal arrangement	The way in which a seal is mounted in the *Seal chamber* and the method of exercising control over the liquid in the *Seal chamber*, viz. *Dead-ended*, *Product recirculation* (see also *API piping plan*).	*Secondary seal*	Seal used to prevent *Leakage* through alternative paths to that between the *Seal faces*. See *Dynamic* and *Static secondary seals*.
Seal chamber	The space in which a mechanical seal is mounted.	*Secondary seal land*	That part of the shaft or *Seal sleeve* in contact with the *Dynamic secondary seal*.
Seal configuration	The design or style of the *Primary seal* (e.g., *Pusher seal*, *Bellows seal*, *Double seal*).	*Shaft sleeve*	A sleeve fitted over the shaft in way of a mechanical seal to provide a wear-resistant and replaceable *Secondary seal land*.
Seal envelope	The external dimensions of a mechanical seal.	*Single seal*	A *Seal arrangement* with only one *Mechanical seal* irrespective of whether other seal types (e.g., *Throttle bush*, lip seal) are included in the *Seal arrangement*.
Seal environment	The physical and chemical conditions prevailing in the *Seal chamber*.		
Seal face(s)	The surfaces of the *Seal ring* and *Seat* in contact with each other.	*Soft face*	*Seal face* manufactured from carbon–graphite or PTFE.
Seal plate	A plate which is bolted to the *Seal chamber* and carries the *Stationary seal member*.	*Solid length*	The axial length of a fully compressed *Mechanical seal*.
		Specific load	*Face load* per unit area of *Sealing interface*.
Seal reference dimension	A reference mark scribed on the shaft to ensure that a *Mechanical seal* is fitted with the correct *Operating length*.	*Spring load*	The load on the *Floating sealing element* exerted by the seal spring(s).
Seal ring	In this publication this term specifically refers to the *Floating seal member* (sprung seal member), though in general usage it can be loosely applied to either of the seal elements. It can be either the *Stationary* or *Rotating seal member*.	*Start-up torque*	The torque absorbed by a *Mechanical seal* on start-up.
		Static secondary seal	Seal used to prevent *Leakage* between assembled parts that are not subject to relative motion in service, e.g., between *Seal sleeve* and shaft, between *Stationary seal member* and *Seal plate*.
Seal size	The maximum diameter of shaft that will pass through the seal, i.e., the diameter of the shaft (or *Shaft sleeve*) to which the *Mechanical seal* is fitted. (It should be noted that alternative definitions based on other dimensions, e.g., *Balance diameter*, are also in current use.)	*Stationary seal*	*Mechanical seal* in which the *Floating seal member* is mounted on the *Seal plate*.
		Stationary seal member	The seal member that is mounted on the *Seal plate*.
		(*Stator*	An alternative term for the *Stationary seal member* of a *Mechanical seal*.)
(*Sealant*	An alternative term for *Barrier fluid*.)	(*Stuffing-box*	An alternative name for *Seal chamber*, carried over from soft-packing technology.)
Sealed fluid	The fluid in the *Seal chamber*.		
Sealed pressure	The fluid pressure in the *Seal chamber*.	*Tandem seal*	*Seal configuration* consisting of a pair of *Mechanical seals* mounted in series (i.e., two *Mechanical seals* sealing in the same direction).
Sealing interface	The contact area between the *Seal ring* and the *Seat*.		
Seat	The axially fixed (unsprung) sealing element. It can be either the *Stationary* or *Rotating seal member*.	(*Thermal stress failure*	Alternative term for *Heat checking*.)
		Throttle bush	A close-fitting bush round the shaft to restrict flow; can be used at the inner end of the *Seal chamber* (*neck bush*) or as an *Auxiliary seal*.
Secondary containment	An arrangement with a chamber on the atmospheric side of a mechanical seal to contain high *Leakage* consequent on failure; this chamber is normally fitted with an *Auxiliary seal*.	*Total closing force*	The sum of the *Hydraulic load* and *Spring load* acting on the

	Floating sealing member to close the *Seal faces*.	*Unbalanced seal*	A *Mechanical seal* in which the *Balance ratio* is greater than or equal to 1.
Toxicity rating	Classification of fluid toxicity defined in N. Irving Sax *Dangerous Properties of Industrial Materials*, 1984.	*'U' ring*	A 'U' section *Dynamic secondary seal*.
		Vent connection	A connection to the *Seal chamber* to allow removal of gas or vapour.
	Toxicity Rating:	*'V' ring*	A 'V' section *Dynamic secondary seal*.

Toxicity Rating:
 0 = no harmful effects under normal conditions.
 1 = short term effects which disappear once exposure removed.
 2 = may produce both short and long term effects, but normally not lethal.
 3 = may cause death or permanent injury even after short exposure to only small quantities.
 U = insufficient data available on humans.

Waviness   Deviation of the *Seal faces* from circumferential flatness. *Waviness* can be present on the *Faces* as manufactured or develop after running.

Wear track   The wear mark of the narrower *Seal face* on the wider one.

Wedge ring   A wedge-section *Dynamic secondary seal*, usually manufactured from PTFE.

APPENDIX 2

NOTATION

In general, the formulae given in the text of this book may be used in any consistent system of units. Units are therefore not shown against the symbols, but dimensions are indicated. However, certain formulae are empirical and must be used in the system of units for which they were developed. The units to be used for these symbols are given below the main table.

Symbol	Description	Dimensions
A	Cross sectional area of seal perpendicular to heat flow	L^2
A_f	Area of sealing interface $\pi(D_o^2 - D_i^2)/4$	L^2
A_h	Hydraulic loading area:	
	internal seal $\pi(D_o^2 - D_b^2)/4$	L^2
	external seal $\pi(D_b^2 - D_i^2)/4$	L^2
b	Radial width of sealing interface $(D_o - D_i)/2$	L
B	Balance ratio:	
	internal seal	
	$(D_o^2 - D_b^2)/(D_o^2 - D_i^2)$	
	external seal	
	$(D_b^2 - D_i^2)/(D_o^2 - D_i^2)$	
c	Specific heat	$L^2T^{-2}\theta^{-1}$
C	Circumference of heat transfer	L
d	Diameter	L
d_s	Outer diameter of rotating seal body	L
d_{sc}	Bore of seal chamber	L
D_b	Seal balance diameter	L
D_h	Outer diameter of stationary secondary seal	L
D_i	Inner diameter of sealing interface	L
D_m	Mean diameter of sealing interface $(D_i + D_o)/2$	L
D_o	Outer diameter of sealing interface	L
D_1	Inner diameter of bellows	L
D_2	Outer diameter of bellows	L
f_t	Closing force used in calculating seal face friction factor	MLT^{-2}
F	Friction force	MLT^{-2}
F_h	Hydraulic closing force $A_f(B\Delta P + p_a)$	MLT^{-2}
F_{hd}	Hydrodynamic force component of interface film	MLT^{-2}
F_h	Hydrostatic force component of interface film $A_f(\beta\Delta p + p_a)$	MLT^{-2}
F_m	Asperity contact load	MLT^{-2}
F_{net}	Net closing force	
	$F_t - F_{hs} - F_{hd}$	MLT^{-2}
	(approximately $F_t - F_{hs}$)	MLT^{-2}
F_s	Spring force	MLT^{-2}
F_t	Total closing force	MLT^{-2}
G	Seal duty parameter $\eta Vb/F_t$	
h	Interface film thickness	L
h_s	Face separation used in calculating leakage	L
h'	Heat transfer coefficient	$MT^{-3}\theta^{-1}$
H	Total power dissipation in seal $H_s + H_{sc}$	ML^2T^{-3}
H_d	Heat dissipated at seal faces	ML^2T^{-3}
H_s	Power generated at sealing interface $(= H_d)$	ML^2T^{-3}
H_{sc}	Fluid friction loss in seal chamber	ML^2T^{-3}
k	Thermal conductivity	$MLT^{-3}\theta^{-1}$
K	Correction factor for seal leakage	
l	Axial length of seal ring heat transfer surface	L
l_c	Circumferential length of seal element	L
l_r	Axial length of rotating seal body	L
L	Life	T
L_{10}	L_{10} life	T
m	Mass	M
M	Friction torque at sealing face μF_t	MLT^{-2}
n	Rotational speed of seal	T^{-1}
p	Sealed pressure	$ML^{-1}T^{-2}$
p_a	Pressure acting on low pressure side of seal	$ML^{-1}T^{-2}$
p_f	Specific net closing force F_{net}/A_f	$ML^{-1}T^{-2}$
p_n	Specific closing force F_t/A_f	$ML^{-1}T^{-2}$
p_s	Specific spring load F_s/A_f	$ML^{-1}T^{-2}$
Q	Flow rate	L^3T^{-1}
r	Radius	L
Re	Reynolds Number (for seal rotation) $\varrho V_s(d_{sc} - d_s)/2\eta$	
R_i	Inner radius of seal ring $D_i/2$	L
R_o	Outer radius of seal ring $D_o/2$	L
T_f	Temperature of interface film in seal	θ
T_p	Temperature of coolant flow in seal chamber	θ
V	Mean relative sliding velocity at seal face $\pi D_m n$	LT^{-1}
V_s	Velocity at outer periphery of rotating seal body $\pi d_s n$	LT^{-1}
β	Pressure distribution factor in interface film	
β_c	Critical value of β at face separation	
β'	Weibull Index	
δp	Difference between specific closing force and pressure drop across seal $\Delta p - p_n$	$ML^{-1}T^{-2}$
Δp	Differential pressure across seal $p - p_a$	$ML^{-1}T^{-2}$
Δp_c	Differential pressure generated by centrifugal effects	$ML^{-1}T^{-2}$
Δp_o	Minimum pressure drop across seal at which face separation can occur	$ML^{-1}T^{-2}$
ΔT	Temperature differential between interface and sealed liquid boiling (bubble) point	θ
ΔT_c	Coolant temperature rise	θ
η	Dynamic viscosity	$ML^{-1}T^{-1}$

161

Symbol	Description	Dimensions
μ	Coefficient of friction	
ϱ	Density	ML^{-3}
ω	Angular velocity	T^{-1}

Symbols used in empirical equations

Symbol	Description	Dimensions
B_2	Impeller tip width (pump)	mm
d_i	Inner diameter of sealing interface	mm
d_o	Outer diameter of sealing interface	mm
D_t	Pump impeller tip diameter	mm
H'	Total power dissipation in seal (empirical)	kW
K_r	Impeller radial load coefficient (pump)	

Symbol	Description	Dimensions
N	Rotational speed of seal	r/min
P	Pressure drop across seal	bar
P_g	Closing force used in calculating leakage $BP + P_s$	bar
P_s	Specific spring load	bar
Q_1	Leakage rate	ml/h
Q'	Empirical leakage rate for 75 mm balanced seal at 3600 r/min	ml/h
S	Empirical gap factor used in calculating seal leakage	bar/s
W_r	Radial load on pump impeller	N
ΔP	Pressure generated across pump impeller	N/mm^2
Φ	Norminal seal size	mm

APPENDIX 3

BIBLIOGRAPHY

A3.1 References to Papers mentioned in the text

Carter, A. D. S., *Mechanical reliability*, 1972 (Macmillan, London).

Rotary mechanical seals. A technical appraisal (Proceedings of Seminar), 1983 (IMechE, London).

Mayer, E., *Axiale Gleitringdichtungen,* 1977, 6th Edn (VDI Verlag, Dusseldorf).

Mayer, E., 1977, *Mechanical Seals*, 1977, 3rd English Edn (translated B. S. Nau) (Butterworth, London).

Sax, N. I., *Dangerous Properties of Industrial Materials,* 1984, 6th Edn. (Van Nostrand Reinhold, New York).

A3.2 Standards referred to in the text

American National Standards Institution
ANSI B73.1 *Specification for horizontal end suction centrifugal pumps.*

American Petroleum Institute
API 610 *Centrifugal pumps for general refinery services.* 6th Edn, January 1981 (7th Edn in preparation).

British Standards Institution
BS 903: Part A6 *Methods of testing vulcanised rubber: determination of compression set after constant strain.*
BS 5257 *Horizontal end suction centrifugal pumps.*
BS 4500: Part 1 *General tolerances and deviations.*

Deutsche Normen
DIN 24 960 *Gleitringdichtungen.* June 1980. (Mechanical seals; cavities; principal dimensions, designation and material codes)

International Standards Institution
ISO 1940: Part 1 *Balance quality of rotating rigid bodies.*
ISO 2858 *End suction centrifugal pumps (16 bar rating) – designation, nominal duty point, and dimensions.*
ISO 3069 *End suction centrifugal pumps – dimensions of cavities for mechanical seals and soft packing.*
ISO 3945 *Criteria for assessing mechanical vibration of machines.*
ISO/DIS 5199 *Technical specifications for centrifugal pumps Class II.*
ISO 5343 *Criteria for evaluating flexible rotor balance.*
ISO 5406 *The mechanical balancing of flexible rotors.*

National Association of Corrosion Engineers
NACE MR-01-75 *Sulfide stress cracking resistant metallic materials for oil field equipment.*

VDI – Richtlinien
VDI 2060 *Beurteilungsmaßstabe fur den Auswuchtzustand rotierender starrer Korper.* (English translation: Peter Peregrinus, Stevenage, Herts.)

A3.3 Further reading

The paper 'Rotary mechanical seals in process duties; an assessment of the state of the art' (B. S. Nau, *Proc. Inst. mech Engrs*, 1985, **199A**, 17–31), which is reproduced at the end of this Appendix, was prepared for the Working Party by Dr Nau as a starting point for their deliberations. It contains an extensive list of the most significant publications in the field of mechanical seals and readers can obtain from it references to particular aspects about which they require further information.

The additional list below indicates some other publications of interest that are not referred to by Nau.

Bloch, H. P., 'Selection strategy for mechanical shaft seals', *Hydrocarbon Processing*, January 1983, **62**, 75–81.

Bloch, E. and **Elliott, H. G.**, 'Mechanical seals in medium pressure steam turbines', *Lubric. Engng.*, 1985, **41**, 653–658.

Bloch, H. P. and **Johnson, D. A.**, 'Downtime prompts upgrading of centrifugal pumps', *Chem. Engng.*, *1986*, **93**, 35–41.

Fern, A. G. and **Nau, B. S.**, *Seals* (Engineering Design Guide, No. 15), 1976 (Design Council, London).

Fujita, T, Matsumato, S. and **Koga, T.**, 'Design and performance of mechanical seals for liquefied gas application', *Lubric. Engng*, 1987, **43**, 440–446.

Flitney, R. K., Nau, B. S. and **Reddy, D.**, *The Seal Users Handbook*, 3rd Edn. (BHRA, Cranfield).

Mechanical Seals Handbook, 1986 (Fluid Sealing Association, USA).

Kojima, K., *et al.*, 'End face seals for high *PV* performance', *Lubric. Engng*, 1985, **41**, 670–674.

Nau, B. S., 'Hydrodynamic lubrication in face seals', *Proc. 3rd Int. Conf. on Fluid Sealing*, 1967, Paper B5 (BHRA, Cranfield).

Pape, J. G., 'Fundamental research on a radial face seal', *ASLE Trans.*, 1968, **11**, 302–309.

Schmidhals, W., *Mechanical seals – causes and analysis of damage*, 1968 (Burgmann GmbH).

Seals Handbook, 1969 (Design Engineering Series, . Morgan Grampian).

Shepherd, K., 'Sealing in the petrochemical industry', *Lubric. Engng*, 1980, **36**, 40–44.

Spores, A. G., 'Controlling pollution with mechanical seals', *Lubric. Engng.*, **31**, 248–253.

Tankus, H., 'End face seals in abrasive service', *Lub. Engng*, 1963, **17, 403–408.**

Will, T. P., 'Effect of seal face width on mechanical seal performance', *Lubric. Engng*, 1984, **40**, 522–527.

A3.4 Trade Literature

Seal manufacturers have available a number of company publications that provide useful information on specific aspects of mechanical seal technology. These publications are normally available on request.

Rotary mechanical seals in process duties: an assessment of the state of the art

B S Nau, BSc, PhD, ARCS, Mem ASME, FIMA
BHRA, Fluid Engineering Centre, Cranfield Bedford

The operating conditions, performance, existing and desirable standards for rotary mechanical seals are discussed. Research needs are reviewed and the need for functional rather than dimensional standards is emphasized.

1 INTRODUCTION

There is a widely held belief that one of the most unreliable components of mechanical equipment in process plant is the mechanical seal. Such seals are used almost universally on centrifugal pumps in process plant in the UK and, in consequence, the national cost of unreliability is probably high in terms of both direct maintenance costs and indirect costs associated with a failure (lost production, capital tied up in stand-by equipment and spare parts, etc.). The object of this paper is to summarize the available facts and this will be done under the following main headings:

1. Operating conditions.
2. Performance in service.
3. Standards.
4. The state of knowledge of mechanisms of seal behaviour.

2 OPERATING CONDITIONS

An indication of the fluids, temperatures and pressures of interest is given by Figs. 1–4 (Flitney and Nau 1976), these represent a sample from 41 sites in a range of industries.

2.1 Pressure

The pressure histograms (Figs. 3 and 4) immediately highlight one area of ignorance, namely the actual operating pressure seen by the seal. In reality this may lie between suction pressure (or even less) and discharge pressure. The pressure in the seal chamber of the pump is determined by a number of factors, including:

(a) pump configuration;
(b) impeller design;
(c) the existence of a seal flushing connection from pump discharge, or suction, to the seal chamber and the flow resistance of this circuit;
(d) the flow resistance of pump wear rings.

Since pressure loading of the seal faces is a key factor in mechanical seal design, the lack of data on operating pressures of the seal is an impediment to good seal design and application engineering.

This paper was presented at a meeting of the Process Industries Division held in London on 15 March 1984. The MS was received on 5 July 1984 and was accepted for publication on 3 September 1984.

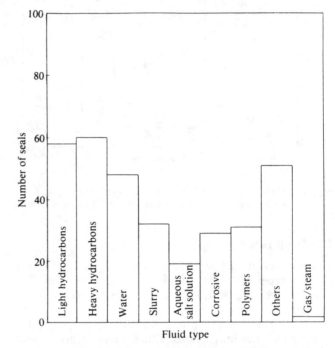

Fig. 1 Histogram illustrating range of fluids sealed by mechanical seals in process plant (Flitney and Nau 1976)

Other pump pressure problems have been pointed out by an Amoco maintenance manager (Nelson 1980). One is pressure pulsation due to recirculation occurring in the pump impeller entry region, particularly in the wide inlets of pumps of high specific speed ($N.Q^{1/2}/\mathrm{NPSH}_{\mathrm{reqd}}^{0.75}$). As a result the seal may experience both fluctuating pressure and axial shaft vibration due to the fluctuating load on the shaft bearings. Nelson also points out that 'the lower 50–60 per cent of many pump operating curves are invalid as operation here results in hydraulic instability'; again this could react adversely on the seal. Pressure surges caused by opening and shutting valves in delivery lines, often at points remote from the pump, may also be a significant factor.

In conclusion, pump pressures cause concern in relation to the seal, both by virtue of the uncertainty of the value seen by the seal and the effect of uncontrolled fluctuations and operation at non-optimum conditions due to changes in plant requirements.

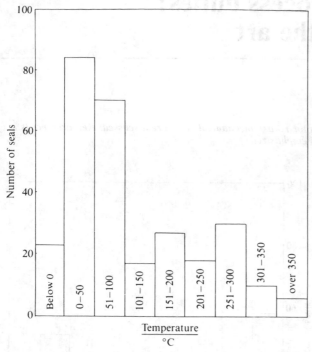

Fig. 2 Typical range of product temperatures sealed by mechanical seals in process plant (Flitney and Nau 1976)

2.2 Temperature

The pumped fluid temperature is not a good indication of the temperature of the fluid in the seal chamber. The latter is often modified by the injection of cold fluid and sometimes a thermal barrier may be fitted between the seal chamber and the pump impeller chamber. Other cooling facilities may also be fitted. Field measurements are plotted in Fig. 5 (Flitney and Nau 1976). These data

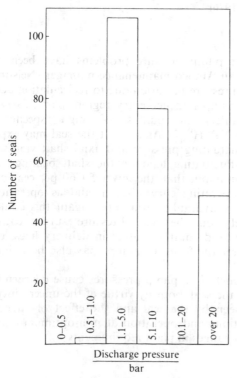

Fig. 3 Typical discharge pressures for centrifugal pumps in process plant (Flitney and Nau 1976)

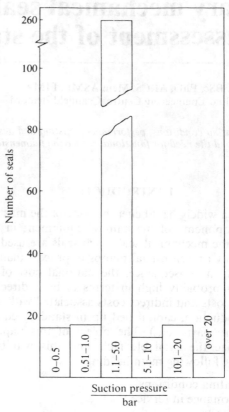

Fig. 4 Typical suction pressures for centrifugal pumps in process plant (Flitney and Nau 1976)

show that below about 80°C pumped product temperature, the seal chamber is typically hotter than the product (due to seal friction), above 80°C the seal chamber is likely to be cooler than the product. Above about 200°C it appears that effective cooling is likely to hold the temperature of the seal chamber to a maximum of about 120°C.

Increasing temperature is reflected in decreasing seal life, as indicated by Fig. 6a (Flitney and Nau 1976).

A considerable, but unknown, proportion of high temperature failures are widely believed to be attributable to:

(a) the additional complexity of auxiliary cooling circuitry;

(b) the accidental loss of cooling.

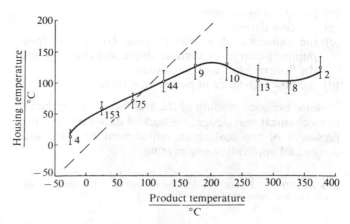

Fig. 5 Seal housing temperatures in relation to product temperatures in a sample of process duties (Flitney and Nau 1976)

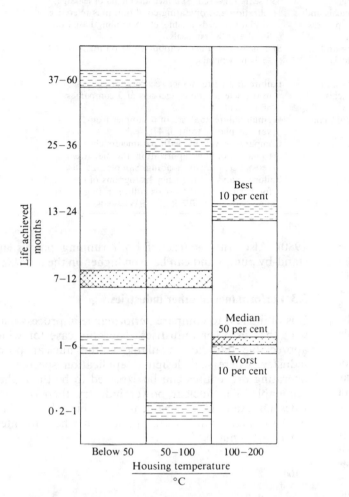

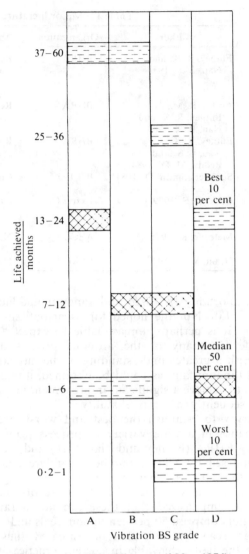

Fig. 6   Effect of (a) seal housing temperature and (b) housing vibration on seal life (Flitney and Nau 1976)

It may therefore be concluded that high temperature failures could be reduced:

(a) if cooling circuitry were more reliable;
(b) if human errors could be reduced;
(c) if seals were able to handle high product temperatures without auxiliary cooling.

3 PERFORMANCE

3.1 Aspects of interest

Two aspects of performance are of primary concern to most users: life and leakage rates. In addition, reliability is important since this enables maintenance to be planned and, also, leakage may be accommodated if the quantity is predictable. Seal life should ideally be some multiple of a normal plant production run. What constitutes 'acceptable' leakage depends very much on the application; thus on a fire pump water leakage is of little consequence, but on a toxic chemical application any leakage at all may be unacceptable whether liquid phase or vapour phase.

In the last few years some performance data have become available as a result of concern for the maintenance costs attributable to pump seals. Table 1 summarizes the published literature.

Leakage to atmosphere is likely to become of increasing concern, with the trend towards stricter environmental control measures.

3.2 Performance in process plant

Data on seal failure rates and costs from one large user in the USA (Exxon) have been published. Buck (1979) studied refinery seals and concluded that 'most seals fail prematurely'. Less than 10 per cent of his sample of 36 failed seals were considered to have worn out. Will (1981) said that this company spent \$15 M per year for maintenance of 14 000 centrifugal pumps and 75 per cent of maintenance costs was attributed to mechanical seal failure. These figures are probably comparable with those from large users in the UK, although data here are less recent. In 1970, BP Chemicals reported that 60 per cent of plant breakdowns were due to mechanical seals and Lankro said that three spares per installed seal were held in stock, implying a high failure level.

A more detailed study of certain groups of seals on centrifugal pumps was recently made at ICI (Summers-Smith 1981) and results were presented as Weibull charts. The L_{10} lives for these are given in Table 2.

The Weibull indices, β, for the samples in Table 2 ranged between 0.7 and 1.1, indicating a mixture of

Table 1 Major literature references dealing with service performance

Authors	Organization	Application area	Notes
Flitney, R. K. and Nau, B. S., 1977	BHRA	Refining, chemicals and other process plant	330 seals surveyed. Site measurements of leakage, vibration and operating condition plus reference to plant records for life, etc. Sampling based on 'best' and 'worst' seals.
Austin, R. M., Flitney, R. K. and Nau, B. S., 1980	BHRA	Refining and chemicals	Site measurements of vapour emission made on 69 seals in 4 plants.
Flitney, R. K., Nau, B. S., and Reddy, M. D., 1984	BHRA	Refining and chemicals	Failure and performance record analysis of seal in pre-selected plant sections at 9 companies. 500+ seals.
Summers-Smith, D., 1981	ICI Ltd	Chemical plant	Weibull failure analyses of 6 samples from various plants, total of 478 seals.
Buck, G. S., 1979	Exxon Co	Oil refining	Compares service life with parameters chosen to measure: film vaporization, thermal stress resistance, pressure loading. Sample size 36.
Metcalfe, R., 1975	AECL	Nuclear power	Failure analysis. Frequency histograms of life for 121 failed and 64 seals still operating.
Grant, W. S., 1977	EPRI	Nuclear power	Failure survey for PWR and BWR seals.

infantile mortality ($\beta = 0.5$) and random mid-life failures ($\beta = 1.0$). Not the hoped for 'wear-out' situation ($\beta \geqslant 3.4$). It is perhaps unreasonable to expect a very long life for many of the arduous process duties, however, if a greater understanding of the mechanisms causing random failures could be obtained, it might be possible to achieve a significant improvement in seal life and consequently in pump reliability.

A survey to establish the best and worst levels of performance in a wide variety of process plants was made by BHRA (Flitney and Nau 1976) and the data were analysed by a large number of different factors. Taking the data as a whole, the life and leakage frequency histograms in Figs. 7 and 8 were obtained. It should be emphasized that these are not a random sample but comprise 50 per cent 'good' seals and 50 per cent 'bad' seals for each plant investigated, thus they indicate the best achievable in current practice at one end of the scale and the worst at the other. The best levels of life and of leakage are impressive and represent a target for the applications where the seals currently fall short of this standard.

The BHRA survey showed that discontinuous operation leads to a notably short running life. The second and third factors, in order, contributing to short life appeared to be vibration (Fig. 6b) and corrosive fluids. Abrasive solids were also believed to be a factor, although the data were confused, since it was not possible to differentiate between 'abrasive' and 'non-abrasive' solids in the sample analysed.

In a follow-on survey, concentrating on hydrocarbon vapour emission by mechanical seals, BHRA found that, even though emission is not visible to the eye, the emission rates of vapour typically lie in the third column of Fig. 8, i.e. the _mass_ flow of vapour is comparable to that of liquid leakage (Austin, Flitney and Nau

1980). Also, this is true of both running pumps and stand-by pumps and can be even higher on the latter.

3.3 Performance in other industries

It is of interest to compare performance in process plant with that in other industries. The only case for which appreciable data are available is the nuclear power industry, where seal design is application specific and operating procedures can be expected to be the highest attainable. The nuclear power industry therefore provides an example of what can be achieved in a high pressure, high temperature application when an intensive effort is made.

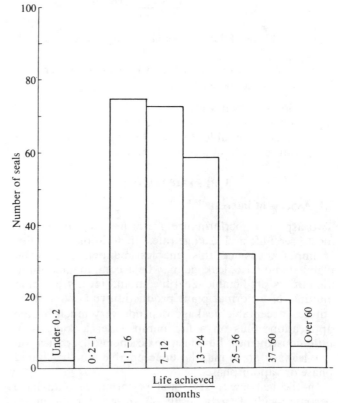

Fig. 7 Seal life distribution from sample of process duties (Flitney and Nau 1976)

Table 2 Seal life in various plants (Summers-Smith, 1981)

Plant type	L_{10} life (days)	Sample size
Nitric acid	1.5–40	c 300
Methanol	43	39
Heat transfer fluid	1.0	38
Light organic fluids	18	71

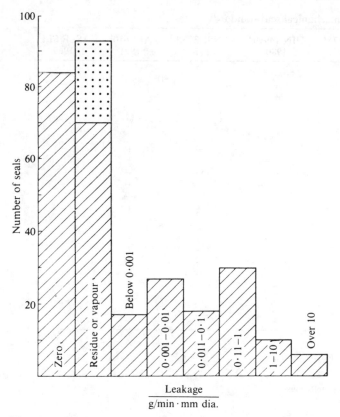

Fig. 8 Seal leakage distribution from sample of process duties (Flitney and Nau 1976)

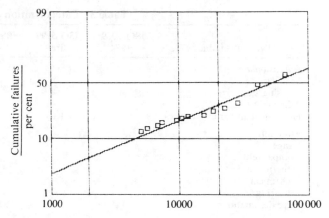

Fig. 9 Weibull chart for nuclear circulator seals, excluding initial failures (L_{10} life 1812 hours, slope $\beta = 0.76$)

(a) improper assembly and venting;
(b) solids in the seal water;
(c) manufacturing errors;
(d) auxiliary system and instrumentation problems;
(e) wear and deterioration of seal faces;
(f) secondary seal problems.

3.4 Conclusions from available performance data

The conclusions which can be drawn from nuclear experience are that:

1. The state of the art has not proved adequate to guarantee the performance of even the most carefully developed mechanical seals for a well defined application.
2. Complexity of auxiliary systems increases the risk of failure.

By comparison with nuclear experience, the present level of reliability of mechanical seals in process plant appears to represent what might be expected, bearing in mind that individual seal designs have to accommodate a whole range of duties; that duties are normally not well defined; and that laid down procedures are possibly less rigorously followed than in nuclear plant. These constraints are unlikely to be relaxed and have to be considered part of the duty specification of a mechanical seal for process plant duty. Given this, however, one must still conclude that seal operating life in process plant is surprisingly short and is certainly less than required. Equally important, reliability is low.

Two published studies are available. Life data from primary circulating pump seals in two Canadian nuclear power plants is given by Metcalfe (1976). Of 169 seals, 85 failed in the first 1000 hours (Metcalfe regarded these as infantile mortality), 48 were still running and the remaining 36 failed after periods of 2000–20 000 hours. The longest running seal had accumulated an impressive 31 000 hours, demonstrating what can be achieved. The very high proportion of initial failures is remarkable but may be due to the size and complexity of these particular seals. The published data lend themselves to Weibull analysis and, if the initial failures are omitted, the result is as shown in Fig. 9. The Weibull Index ($\beta = 0.76$) still indicates premature failures rather than wear-out although the lives are much better than for the process plant data above (although the infantile mortalities were included in that analysis): the L_{10} life is 1800 hours. It is interesting that even in a closely controlled application with purpose-designed seals there are still 'mid-life' failures. Among the factors reported as contributing to circulator seal failures, Metcalfe lists:

(a) excessive face waviness;
(b) creep or relaxation of seal structural materials;
(c) erosion and corrosion;
(d) abrasion;
(e) surface failure (cavitation and thermal crazing);
(f) fracture;
(g) secondary seal problems.

Experience with nuclear power plant circulator seals in the USA has been reviewed in a report published by EPRI (Grant 1977). In a two year period 38 failures occurred in 49 stations. As with AECL, initial failures were predominant. These were attributed to:

4 MECHANICAL SEAL RELATED STANDARDS

4.1 Existing standards

The current situation is summarized in Table 3. This indicates the existence, or otherwise, of standards falling in various categories. It can be seen that in most cases these standards are primarily pump standards and only secondarily seal standards. The main exception is DIN 24960, which gives seal envelope dimensions as well as those of the housing. In addition it gives surface finish requirements and presents a type designation coding system which includes material coding.

The BS 5275 and ISO 3069 define the diametral

Mechanical Seal Practice for Improved Performance

Table 3 Categorization of mechanical seal standards

Type of standard	ISO 3069[1] 1974	ISO 5199[2] draft	BS 5275[3] 1975	DIN 24960[4] 1980	ANSI B73.1[5] 1974	API 610[6] 1981	ASLE SPI[7] 1969
Dimensional							
Seal sizes	−	−	−	+	−	−	−
Housing/shaft sizes	+	−	+	+	+	−	−
Alignment tolerances	−	(+)	−	−	+	−	−
Housing/shaft surface finishes	−	(+)	−	+	−	−	−
Materials							
Usage	−	−	−	−	−	+	−
Compatibility:							
thermal	−	−	−	−	−	+	−
chemical	−	−	−	−	−	+	−
bearing	−	−	−	−	−	−	−
Type designations	−	−	−	+	−	+	−
Functional							
Balance	−	−	−	−	−	−	−
Seal chamber pressure	−	−	−	−	−	+	−
Systems	−	(+)	−	−	−	+	−
Testing	−	−	−	−	−	−	−
Performance	−	−	−	−	−	−	−
Terminology	−	(+)	−	−	−	+	+
Classification	−	−	−	+	−	+	−

Notes
1. End suction centrifugal pumps: dimensions of cavities for mechanical seal and for soft packing.
2. Technical specification for centrifugal pumps: Class II.
3. Specification for horizontal end-suction centrifugal pumps (16 bar).
4. Gleitringdichtungen: Wellendichtingsraum Hauptmasse, Bezeichning und Werkstoffschlussel.
5. Specifications for horizontal, end suction centrifugal pumps for chemical process.
6. Centrifugal pumps for general refinery services.
7. A glossary of seal terms.

dimensions of the recess into which the seal fits but not other dimensions of the seal, unlike the DIN standard. However, all three of these standards, and ANSI B73.1, essentially define a housing to suit a *packed gland*. They do not pay regard to the functional requirements of a mechanical seal. The result is that the mechanical seal and its heat transfer to the sealed fluid, cannot be designed for optimal performance but must be subordinated to the dimensional requirements of soft packing. Neither is the control of vapour and solids around the seal addressed. *The optimization of housings is an area requiring experimental investigation prior to the definition of an improved standard.*

An important parameter in determining seal performance and requirements is the seal chamber pressure. At present there is little information on pressure levels in current pumps nor of the extent to which this varies in service. This pressure is determined partly by pump design features and partly by seal auxiliary circuitry (for injection, etc.). Thus both pump manufacturer and seal manufacturer contribute to the actual (unknown) value. API 610 specifies 25 lbf/in² above suction pressure, while 50 lbf/in² above suction is specified by some seal manufacturers. There is a need to obtain quantitative information on the current situation with a view to introducing a degree of standardization and predictability into seal chamber pressures.

Related to the question of sealed pressures is the subject of seal 'balance', i.e. the degree to which the closing force on the seal is offset by adjusting the dimensions of the various areas exposed to fluid pressure. Until recently API 610 specified discharge pressures above which 'balanced seals' were required (these are not specified in the 1981 edition) and this is common practice in manufacturer's literature. The point which must be made here is that 'balance' is a continuous variable and the amount of load offset can be varied between zero and 100 per cent or even wider limits. There is a need for greater understanding of the role of pressure. Why, for example, if unbalanced seals (balance ratio ~1.1) are limited to a pressure of 10 bar, should it be possible to operate at pressures *above* 18 bar (= 1.1 × 10/0.6) with seals having a balance ratio of 0.6 (about the maximum used in practice)?

The degree of balance required for a given seal actually depends on the sealed pressure. No detailed published information is available on this aspect of seal design. It is suspected that this is *not* a design feature which is rigorously tailored to actual service pressures expected for particular duties but rather that a compromise value is used. In view of the uncertainty as regards seal chamber pressures this tolerance may be justifiable, however there is a very strong case for taking steps first to ensure that the seal pressure is clearly definable and, second, to match the degree of balance to the expected pressure range.

API 610 includes a classification system with material coding but is primarily concerned with the sealing system as a whole, giving throttle bush details and diagrams of various multi-seal systems with their auxiliary piping circuitry. A conspicuous omission is the absence of recommendations on *when to use* these alternative systems. The draft ISO 5199 covers much of the same ground as API 610, it contains similar diagrams and similarly lacks guidance on their use. The missing information is available, e.g. in seal manufacturers' literature and could be incorporated without need for research. This area requires input from seal manufacturers.

4.2 Possible future standards

There are several areas where standards could serve a useful purpose but they are not available at present.

The need for revision of housing dimension standards has been referred to above, also the need for advice on the use of the various auxiliary sealing system options.

In relation to seal materials there is a need for definitive advice on: (*i*) chemical compatibility, in particular, under the rubbing conditions of face contact; (*ii*) thermal limits; and (*iii*) sliding face load bearing limits (e.g. *PV* values). The information required to meet (*i*) and (*ii*) is largely available and is listed with varying degrees of consistency in seal manufacturers' literature. The third item, load limits, requires further experimental investigation which should lead to a definition of test piece geometry and test procedure, and then proceed to the measurement of reference values for seal face materials.

An area where standards are notably lacking is that of standardized performance testing. This requires experimental work to define the test rig and test procedure. Beyond a basic test there is scope for the definition of de-rating factors to take account of deviation of operating conditions from the standard test conditions.

4.3 Summary of research tasks relating to improved standards

1. Develop a mechanical seal test procedure for defined classes of application.
2. Develop a face material test procedure and acquire *PV*-limit data for seal face materials.
3. Define degrees of balance required in relation to sealed pressure.
4. Optimize pump seal chamber design to ensure effective heat transfer and lubrication of the seal and control of vapour and free solids.
5. Establish existing seal chamber pressure levels and variability.

5 BEHAVIOUR MECHANISMS

5.1 Tribological background

The central problem in mechanical seal research is understanding of the behaviour of the sliding interface and all the factors affecting this. The tribological processes involved are very dependent on the nature of the sealed and ambient fluids; the latter is often, but by no means always, air.

Operating conditions at the interface are quite severe. Speeds are typically 2–12 m/s, often higher, and the unit load is typically 0.2–2 MPa. The sealed fluid is often of low viscosity, typically aqueous or a hydrocarbon, but can be any fluid which is pumped; it may be above its atmospheric boiling point and can contain suspended solids.

It is generally accepted that the sealed fluid penetrates the interface and is responsible for supporting most if not all of the load. Support can be provided in part by hydrostatic pressure, the sealed pressure falling off to ambient across the interface in a way which depends on details of any radial taper. In addition, hydrodynamic pressure can be generated by the inter-

action of circumferential viscous shear with variations in face separation, due to residual face waviness, and this too can provide significant load support under suitable conditions. The hydrodynamic mechanism is analogous to that in journal bearings. Experimental data indicate that the fluid film between the sliding faces has a thickness of the order of 0.001 mm.

Tables 4 and 5 classify much of the significant literature in this field.

5.2 Face waviness

Residual waviness of lapped surfaces can be as low as 0.0001 mm PTP (half a 'light band') but this is still significant in relation to the fluid film thickness for the purposes of hydrodynamic pressure generation. More importantly, however, it is found that waviness normally increases in use, rapidly reaching levels of up to 0.010 mm PTP. Such values are measured in the unloaded state and it must be assumed that the actual value under load will be reduced by flexure of the seal ring.

The most detailed analysis of induced waviness was that carried out at the University of New Mexico (Lebeck 1976). He assessed a variety of possible sources, carrying out detailed elastic analyses to assess probable magnitudes of different mechanisms. The mechanisms considered were of three main types:

(a) non-axisymmetric material properties;
(b) non-axisymmetric geometric features;
(c) non-axisymmetric loads.

The conclusion was that the largest of these effects is that due to non-axisymmetric drive torque loads. A limitation of this study is that it refers to floating carbon designs, whereas many mechanical seals have a fixed carbon and the driven ring is a less flexible high modulus material, such as stainless steel or silicon carbide. Furthermore, induced waviness is known to occur even when the drive force is applied uniformly (Nau 1963).

A mechanism requiring investigation is direct elastohydrodynamic (EHD) behaviour in conjunction with local surface wear. This could be a source of induced waviness since locally hydrodynamic pressures can be quite high in relation to the elastic modulus of the face. The EHD effect without wear was subject to preliminary analysis some years ago at the University of Lyon (Lohou 1970) and elasto-hydrostatic effects at BHRA (Nau and Turnbull 1961). These studies require further development to be directly useful in establishing relevance to induced waviness and film load capacity.

To summarize, induced waviness is large compared with initial waviness. The mechanism of formation of induced waviness has not been fully established. It is not known what initial and induced waviness levels represent optimum targets in seal design.

5.3 Structural deflections

Seal face deflections affect the interfacial film geometry and can therefore have a marked effect on seal performance. 'Intrinsic' deflections may be grouped under the following headings:

Table 4 Steady state theoretical analyses

Hydrodynamic and hydrostatic analysis	No distortions		Elastic analysis			
			Coupled axisymmetric distortions (hydraulic and thermal)		Non-axisymmetric distortions	
	Plane faces	Wavy faces	Plane faces	Wavy faces	Uncoupled	Coupled (EHL)
Single-phase film only	Various workers 6 Zuk, 1976 7 Etsion, 1981 9 Nau, 1981 9	Various workers 6 Iny et al, 1971 4	Nau and Turnbull, 1961 1 Fisher, 1961 3 Watson, 1965 3 Cheng, 1967 etc. 1, 3 Bupara et al, 1967 3, 7 O'Donoghue, 1966, 1968, 1969 1 Metcalfe 1976, 1980 3	Nau 1978 4	Iny 1971a 1, 2	Lohou 1970 4, 1
Two-phase film — Isothermal (cavitation)	Anno, 1967 8 Kojabashian, 1967 8	Findlay, 1967 4 Nau et al, 1978, 1980 4 Pape, 1969 4 Ikeuchi and Mori, 1980 4 Ruddy et al, 1982a, b 5 Lebeck, 1977, 1981 4				
Temperature dependent	Hughes, 1980 Lebeck, 1980b			Lebeck 1976, 1980a. c 2, 3, 4		Lebeck 1976, 1980 2
Asperity contact and mixed modes	Teale and Lebeck, 1980 Lebeck, 1978, 1980	Lebeck, 1976a, b, 1979 Burton et al, 1976a, b, c, 1978, 1979, 1980 Kennedy, 1981				

Notes

1. local compressive deflection; 2. circumferential bending; 3. coning; 4. one face wavy; 5. both faces wavy; 6. liquid phase; 7. gas; 8. micro-asperity model; 9. vibration effects

Table 5 Experimental studies

('Low' sealed pressure < 1MPa, 'Low' face load < 1MPa) Organization	Authors	Year of publication	Instrumentation	Visualization	Leakage	Friction/power	Topology	Life	Wear	Damage
1 LIQUID FILM										
1.1 Low sealed pressure										
1.1.1 Low face load										
BHRA, UK	Denny	1961	+		+	+	+			
BHRA, UK	Nau	1963, 1966, 1967, 1969, 1979	+	+	+	+	+			
BHRA, UK	Flitney, Nau and Reddy	1984			+	+	+		+	+
ICI Ltd, UK	Summers-Smith	1961, 1981			+	+	+	+		
Nippon Oil Seal, Japan	Ishiwata and Hirabayashi	1961, 1967	+		+	+			+	
Pure Carbon Co, USA	Paxton *et al*	1961, 1973, 1977, 1980				+			+	
Pure Carbon Co, USA	Strugala	1972								+
CEGB, UK	Batch and Iny	1964, 1971a, b, c, d	+	+	+		+			
Delft University, Holland	Pape	1969	+		+	+				
Lyon University, France	Lohou	1970, 1973, 1978	+	+		+				
Lyon University, France	Haardt	1975	+							
University New Mexico, USA	Lebeck	1976, 1981			+	+	+		+	
Northwestern University, USA	Banerjee and Burton	1978, 1979	+							
Sealol Inc, USA	Labus	1980				+				
Admiralty MTE, UK	Tribe	1983								+
General Motors Research Laboratories, USA	Symons	1974					+			+
Crane Packing Co, USA	Kanasaki and Schoenherr	1975			+				+	
Crane Packing Co, USA	Dziedzic	1980			+		+		+	
Crane Packing Co, USA	Piehn	1965							+	
Crane Packing Co, USA	Trytek	1973							+	
ITT, USA	Janetz	1965					+		+	
HER Institute, USSR	Golubiev *et al*	1975							+	
Kyushu University, Japan	Hirano *et al*	1978		+	+	+				
NMI, Holland	Pijcke and Vries	1978			+	+		+		+
Napier Ltd, UK	Moores and Marsh	1961			+				+	
1.1.2 High face load										
Burgmann, West Germany	Mayer	1961			+	+			+	
1.2 High sealed pressure										
1.2.2 High face load										
Burgmann, West Germany	Mayer	1961			+	+			+	
BHRA, UK	Fisher and Nau	1967			+					
University of New Mexico	Lebeck *et al*	1981			+	+	+			
US Navy, USA	Kojabashian and Richardson	1967					+			
US Navy, USA	Dray; Karpe	1981					+			+
AECL, Canada	Watson	1965, 1973			+		+			+
AECL, Canada	Metcalfe *et al*	1973, 1982	+		+	+				
2 VAPOUR FILM										
2.1 Low sealed pressure										
2.1.1 Low face load										
MTI, USA	Orcutt	1969	+	+	+	+				
Crane Packing, USA	Netzel	1980a, b, 1982					+			+
Exxon, USA	Will	1981	+			+				
Esso, UK	Barnard and Weir	1978					+	+		+
BHRA, US	Nau *et al*	1963, 1969, 1980		+						
2.1.2 High face load										
Burgmann, West Germany	Mayer	1961			+	+			+	
2.2 High sealed pressure										
2.2.2 High face load										
BHRA, UK	Nau	1982			+	+				
Flexibox Ltd, UK	Taylor and Jones	1983				+	+			
3 DRY CONTACT										
3.1 Low sealed pressure										
3.1.1 Low face load										
RAE, UK	Lancaster *et al*	1963, 1975, 1978				+			+	
NASA Lewis, USA	Johnson, Swikert and Bailey	1956				+			+	
NASA Lewis, USA	Bisson, Johnson and Anderson	1957							+	

1. Axisymmetric deflections: these cause the interface to become coned and thereby affect the hydrostatic component of film pressure. These deflections can be caused by:
   (a) hydraulic moments acting on the seal rings;
   (b) thermal expansion and thermal stresses;
   (c) compressive deflection of the seal faces by the fluid film pressure.

2. Non-axisymmetric deflections: these can be caused by interaction of hydrodynamic film pressures with the sealing rings, in particular by:
   (a) 'beam bending' due to uniform loading on one end of a seal ring and non-uniform on the other, due to hydrodynamic pressure;
   (b) local compression due to hydrodynamic pressure, analogous to 1c above.

In addition to these intrinsic effects there are others which are due to design or manufacturing effects:

1. Deflections due to non-uniform drive forces or non-axisymmetric seal ring geometry.
2. Deflections due to imperfection in the mounting arrangements of the seal, for instance pump housing distortion or misalignment.
3. Deflections due to non-uniformity of materials.

As indicated in Table 4, intrinsic axisymmetric effects have been widely studied compared with non-axisymmetric and non-intrinsic effects. Lebeck (1976a) has however carried out detailed analyses of a range of possible causes of effects relating to 2a.

There has been relatively little work on the thermal effects, which require detailed knowledge of heat generation and heat transfer processes in the seal, both areas where there is a lack of good experimental data.

The other main area requiring study is that of EHD behaviour, particularly for carbon face seals, since these involve a low modulus material (about 10 per cent of steel) and comprise the majority of mechanical seals in service. Computed hydrodynamic pressures for rigid wavy surfaces indicate that pressure of sufficient magnitude to cause significant local face compression can occur, especially during starting and stopping.

5.4 Interface lubrication

It was indicated above that fluid penetrates the sealing interface. When this fluid is a liquid phase the hydrodynamic effects referred to can be a major factor in supporting the load on the interface and together with the hydrostatic pressure may actually account for the entire load support. However, if the fluid vaporizes in the interface, then the film thickness must reduce to generate sufficient load capacity. In this situation it is conceivable that a significant degree of asperity contact may occur between the two faces.

There is direct experimental evidence for the existence of fully developed liquid films in mechanical seals (Denny 1961, Nau 1963) also for essentially gaseous films (Nau 1963, Orcutt 1969). There is also indirect evidence for pure hydrodynamic and mixed boundary/hydrodynamic films, this is from friction data plotted as friction coefficient versus duty parameter (i.e. a Sommerfeld or Gumbel plot) (Summers-Smith 1961, Ishiwata and Hirabayashi 1961, Pape 1969, *et al*).

There is also indirect evidence for a degree of asperity contact under certain conditions, this is provided by observed high power dissipation levels of seals on test rigs (e.g. Fig. 10). The high levels sometimes observed cannot be credibly accounted for by purely viscous friction but could be produced by asperity shear over about 2 per cent of the interface area. These high power levels appear to be associated with low leakage rates and film vaporization. In one instance in particular, at BHRA, Reddy observed that the sealed liquid was vaporizing on the high pressure side of the seal, water hardness deposits accumulating there (Nau 1982). Again, it is quite common to observe seals operating with no liquid phase leakage and, whilst it is possible that the leakage evaporates at the edge of the interface it is equally possible that the film is largely in the vapour phase. Barnard and Weir (1978) examined used seals, failed and unfailed, and they noticed radial zoning which was thought to indicate phase changes in the film as the fluid leaked radially. Will (1981) has made interesting temperature measurements which indicate a seal

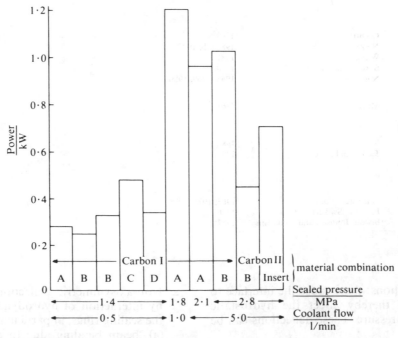

Fig. 10 Examples of power dissipation for a water seal, carbon versus tungsten carbide at various pressures and coolant conditions. The codes A . . . D denote tungsten carbide coating grades; for one test a solid insert of tungsten carbide was used. Two carbon grades were used (Nau 1982)

operating around the transition from liquid film to vapour. Again, measured wear rates of seal carbons of the order of 1 μm/h also suggest some face contact (Paxton *et al*).

To summarize the present state of knowledge, both liquid films and vapour films can occur between mechanical seal faces. In the latter case, at least, asperity contact must also occur in order to account for the observed power dissipation. In the latter case, too, the possibility must be considered that asperities carry a significant proportion of the load on the interface. What is not yet clear is the extent to which actual seals in process plant fall into these two categories of operating mode. Certainly it seems that seals for 'volatile' and 'non-volatile' liquids may behave quite differently.

5.5 Wear

Carbon-based materials form one seal face in the majority of mechanical seal designs and most seal face wear data relate to carbons. It appears that a steady carbon wear rate is reached within the first few hours of running, although friction and temperature continue to fall for up to about 100 hours (Paxton and Shobert 1961). Under normal conditions the hard counterface usually wears very little.

Typical wear of a carbon seal ring with clean water is of the order of 4 mm/10^5 km (Paxton and Shobert 1961 and BHRA data) but there is a wide spread depending on the duty, values as low as 0.004 mm/10^5 km have been reported (Netzel 1982), for a hydrocarbon duty. In plant service, grooving of the carbon is observed in two out of three seals examined (BHRA unpublished).

Carbon wear mechanisms under dry sliding conditions have been studied extensively at RAE (Lancaster 1963–78). Most of this work was with crossed cylinders rather than the continuous contact configuration of a mechanical seal, caution is therefore needed in applying these data to seals. However, the thermally induced high wear transition observed by Lancaster appears to have a parallel in observations with seals by Paxton and Shobert (1961). In this connection the marked reduction in wear produced by various 'contaminants', reported by Lancaster, may be significant. Lancaster has presented evidence for the existence of a compacted layer of carbon debris on sliding carbon surfaces, he explains the transition to high wear in terms of the break-up of this layer and proposes that the layer is stabilized by molecular forces which can be disrupted by high temperatures or made more stable by the 'contaminants' referred to above.

The relevance of Lancaster's carbon layer to mechanical seal conditions requires investigation and could be the key to 'random' fluctuations in seal performance and/or premature failures.

While hard faces do not wear in the sense of a loss of axial length of the seal face, circumferential grooves much deeper than the thickness of the face separating film are commonly observed on seals removed from service, even on those that are still performing satisfactorily, particularly at the outer and inner diameters of contact with the 'soft' face. The mechanism of formation and the significance of these grooves is not fully understood and requires further investigation.

5.6 *PV* values and face compatibility

The compatibility of mating face materials in the sliding contact of seals is of considerable interest. There is a tradition of using a *PV* factor to define the maximum load, *P*, and speed, *V*, combination sustainable by a material pair. Usually additional constraints are also imposed, in the form of a maximum load and a maximum speed. The basis of the *PV* factor is that, given a constant coefficient of friction, the product *PV* is proportional to the heat generated at the sliding interface. The limiting *PV* for a material combination is therefore a measure of the ability of the two materials to limit the amount of heat produced, and to conduct this heat away whilst maintaining a surface temperature below the softening point of the materials and temperature gradients below those which cause surface crazing due to thermal stresses.

While not generally applicable to pump applications, there is also some evidence (e.g. from low speed agitator seals) that there are minimum values of *P* and *V* (if not for *PV*) for satisfactory operation. This is primarily a matter of film formation, but it does have an impact on materials if the film breaks down.

In addition, there is a need for greater understanding of the role of such physical properties as thermal conductivity, coefficient of thermal expansion, thermal shock resistance and wettability. These are factors which may be reflected in maximum *PV* factors.

At present, there is no standardization of test piece geometry, nor of the procedure by which *PV* limits are measured, hence, in general, values cannot be compared with each other or applied to seals of differing cross-section. To achieve standard values it would be necessary, among other things, to standardize the geometry to control thermal distortion which could cause non-uniform loading; to standardize the surface topology to control hydrodynamic effects; and to standardize ambient fluid flow conditions to ensure that heat transfer from the seal to the fluid is adequately controlled.

If standard values of *PV* were available it would be possible to assess the extent to which a particular seal design achieved the potential *PV* limit and if it did not achieve this value the shortfall would be an indication of the seal's potential for improvement.

5.7 Power dissipation

To analyse or design a mechanical seal it is necessary to predict the temperature distribution within the seal, in order to calculate self-induced thermal deflections of the sealing faces and film temperatures. An accurate knowledge of the heat dissipation in the interface is necessary to do this. Unfortunately, such information is not generally available. The data provided in commercial literature vary widely and appear to be of dubious value for this purpose. Work at Sealol Inc has provided useful data (Labus 1980) but is nevertheless limited in scope.

It is possible that a standard power dissipation test could be combined with a standard *PV* test to provide basic data for design and analysis purposes.

5.8 Vibration effects

Passing reference to some sources of vibration and the effects of vibration on seal performance has been made

Table 6   Vibration sources affecting mechanical seals

Origin	Type	Possible source
Self-induced	Angular misalignment of faces to axis of rotation: both faces misaligned one face misaligned, other wavy	Manufacturing tolerances of pump, seal or both
	Both faces wavy	Manufacturing tolerances of seal
	Dynamic out-of-balance of seal rotor	Non-axisymmetric design features, e.g. drive slots
External	Shaft vibration: axial radial	Pump flow mismatch Pump flow mismatch; sleeve eccentric
	angular torsional	Diesel drive, warped shaft, etc.
	Housing vibration: locally excited	Via pipework or foundations
	remote excitation	As above

above (Sections 2.1, 3.2). A more comprehensive discussion is given in Nau 1981a, and 1981b. A summary of the sources of vibration to which mechanical seals are exposed is given in Table 6 and a summary of the literature in Table 7.

5.9 Failure mechanisms

The transition from a stable running condition, with the faces largely separated by a fluid film with low or no wear, to breakdown of the fluid film and a change to a high wear or surface damage condition, is an important stage in the life of the seal, though it may not be the point of failure observed by the user and indicated by unacceptable leakage. This complicates the investigation of seal failures on operating plant.

The direct mechanism of failure referred to above probably underlies many indirect failure modes too. Indirect modes tend to be better understood, at least in relation to their initial causes. In this category can be mentioned failures attributable to the secondary seal, or springs; failure due to abrasives in the sealed fluid; solids deposition; material/fluid incompatibility; loss of fluid; excessive vibration, etc. Many failures occur very

Table 7   A classified summary of the literature of mechanical seal vibration studies

Fluid type	Seal type	Study	Subject	Authors	Date	Corporate reference
Gas	Bellows mechanical seal	Experimental	Observation and elimination of instability	Strom *et al*	1969	NASA
		Analytical	Seal ring dynamics	Hill	1969	North Carolina State University
	Shrouded-step seal (circumferentially flexible)	Analytical Analytical Analytical	Face tracking Face tracking Stiffness	Winn *et al* Ludwig Cheng *et al*	1968 1978 etc 1968	MTI NASA MTI
	Spiral-groove seal	Analytical Analytical	Face tracking Stiffness	Kupperman Gabriel	1978 1978	Crane (USA) Crane (USA)
	Spiral-groove seal with hydrostatic orifices	Analytical	Stiffness	Cheng *et al*	1968	MTI
Liquid	O-ring mechanical seal	Experimental	Response to radial vibration of shaft	Rowles and Nau	1981	BHRA
			Air ingestion and face angular alignment	Kaneta and Fukuhori Nau Nau	1978 1963 1968	Kyushu University BHRA BHRA
			Diametral compressive dynamic stiffness and damping of rubber O-rings	Smalley *et al* Rowles and Nau	1977 1978/1981	MTI BHRA
			Diametral rocking stiffness of rubber O-rings	Metcalfe	1978	AECL
			Axial shear stiffness of rubber O-ring (large displacements)	Kittner and Metcalfe	1980	AECL
			Response to shaft run-out	Sehnal *et al*	1983	Crane (USA)
		Analytical	Integrated dynamic response analysis including film hydro-dynamics/hydrostatics, O-ring mount and ring inertia	Rowles and Nau	1981	BHRA
			Tracking of misaligned faces	Haardt Etsion	1975 1980, 1981	University of Lyon NASA
			Magnitudes of stiffness and damping forces acting on floating seal ring	Rowles and Nau	1978	BHRA
	Coned hydrostatic seal	Analytical	Stiffness Stability	Nau and Turnbull Metcalfe	1961 1980/81	BHRA AECL
		Experimental	Whirl instability Response to shaft run-out	Metcalfe Sehnal *et al*	1980, 1982 1983	AECL Crane (USA)
	Orifice-fed hydrostatic pocket seal	Analytical	Angular stiffness	Fisher	1963	BHRA

early and are attributable to handling/installation problems, in most of such cases the problems are not of a fundamental nature. However, it could be argued that a really good seal design would not be susceptible to such problems. The detailed failure mechanisms of different face material combinations and their ability to withstand transient breakdown conditions requires further study. One failure mode which has been studied fairly extensively and is now reasonably well understood is thermal crazing. There is evidence that this is due to compressive yield in the surface due to excessive thermal expansion, followed by tensile fracture as the surface cools (Kennedy 1981).

5.10 The literature

Having discussed some of the important topics relating to the behaviour of mechanical seals it is appropriate to summarize the effort which has gone into the different areas. For this purpose, particularly significant research publications have been classified in Table 4 (analytical studies) and Table 5 (experimental studies). The full references for items in these tables are given in the bibliography. It can be seen from these tables that most effort has been expanded in the less complex areas, e.g. liquid films with no or axisymmetric distortions. Comprehensive studies of temperature effects are lacking. Also, asperity contact studies tend to be theoretical or with rather indirect experimental support from the viewpoint of industrial mechanical seal applications, there is clearly scope for work in this area if only to establish where and when asperity contact must be considered.

6 GENERAL CONCLUSIONS

Currently, information on actual operating conditions on centrifugal pumps is imprecise, particularly in respect of pressure at the seal. An additional element of uncertainty may also arise in particular installations due to operational deviations both of plant duty or the seal auxiliary systems used to control seal chamber conditions by recirculation, injection, etc; this lack of information and control is likely to be a contributory factor to reduced seal life.

Large users agree in attributing a large proportion of process plant maintenance costs to mechanical seal failure. Seal lives are also considered to fall short of what is required in very many cases. The fact that a few seals exhibit impressively long life indicates that acceptable life should be possible for a larger proportion of seals than at present. The reasons for unpredictable premature failure can be many but appear to include some basic factors not yet fully understood. The nuclear industry in Canada achieves longer seal lives on circulator seals but still experiences premature failures. This seems to substantiate the above conclusion.

The situation as regards standards relating to mechanical seals appears in need of improvement:

1. Dimensional housing standards have been imposed without regard to the operational requirements of mechanical seals, such as heat transfer and elimination of vapour and solids.
2. Whilst 'balanced' seals are sometimes specified for pressures above certain values, no attention is given to the quantitative nature of this parameter.
3. Standards do not specify when, or what type of, auxiliary circuitry is required by a seal nor specify the performance required of this. In particular, pressure at the seal is not adequately controlled.
4. Standards do not exist for performance testing mechanical seals nor for measuring compatability of materials, in particular PV values are quoted but there is no standard procedure for their measurement.

Research into sliding interface behaviour has established the possibility of various operating regimes such as: 'liquid film', 'asperity contact' and 'vapour phase'; what is lacking is clear information on the limits of each in terms of practical seal operating conditions. The origin and significance of induced waviness is still not fully understood. The analysis of interactions between structural deflections and the interface are fairly well developed except in the area of possible elastohydrodynamics due to local compressive deformation of (particularly) low modulus faces. Prediction of thermal deflections is hampered by inadequate interface heat dissipation data.

Face wear mechanisms have only been studied in depth using test pieces which do not simulate seal interface conditions adequately. There is evidence from such work that a protective layer of carbon particles can be formed on the sliding surface of carbon, its existence depending on molecular bonds which can be disrupted by elevated local temperatures or enhanced by chemical contaminants. The possibility should be investigated that disruption of such a layer on mechanical seal faces could be responsible for unpredictable premature failure of seals.

BIBLIOGRAPHY

Anno, J. N., Walowit, J. A. and **Allen, C. M.** Microasperity lubrication. *3rd Int. Conf. on Fluid Sealing*, Paper E2, BHRA, 1967.

Austin, R. M., Flitney, R. K. and **Nau, B. S.** Rotary seal vapour emission. BHRA Report RR1616, Sept. 1980.

Banerjee, B. N. and **Burton, R. A.** Thermoelastic instability in lubricated sliding between solid surfaces. *Nature*, June 1976b, **261**, 399–400.

Banerjee, B. N. and **Burton, R. A.** An instability for parallel sliding of solid surfaces separated by a viscous liquid film. *Trans. ASME, J. Lubric. Tech.*, Jan. 1976a, 157–165.

Banerjee, B. N. and **Burton, R. A.** Experimental studies on thermoelastic effects in hydrodynamically lubricated face seals. *Trans. ASME, Lubric. Tech.* July 1979, **101**, 275–282.

Barnard, P. C. and **Weir, R. S. L.** A theory for mechanical seal face thermodynamics. *8th Int. Conf. on Fluid Sealing*, Paper H1, BHRA, 1978.

Batch, B. A. and **Iny, E. H.** Pressure generation in radial-face seals. *2nd Int. Conf. on Fluid Sealing*, Paper F4, BHRA, 1964.

Bisson, E. E., Johnson, R. L. and **Anderson, W. J.** On friction and lubrication at temperatures to 1000°F with particular reference to graphite. ASME Preprint No. 57-Lub-1, 1957.

Buck, G. S. A methodology for design and application of mechanical seals. *Trans. ASLE*, 1980, **23**(3), 244–252.

Bupara, S. S., Walowit, J. A. and **Allen, C. M.** Gas lubrication and distortion of high pressure mainshaft seals for compressors. *3rd Int. Conf. on Fluid Sealing*, Paper B3, BHRA, 1967.

Burton, R. A. *et al.* The influence of thermoelastic instabilities on face seal contact. *Proc. 1976 Tribology Convention* 1978, pp 53–57 (IMechE, London).

Cheng, H. S. and **Snapp, R. B.** A study of the radial film and pressure distribution of high pressure face seals. *3rd Int. Conf. on Fluid Sealing*, Paper E3, BHRA, 1967.

Cheng, H. S., Castelli, V. and **Chow, C. Y.** Performance characteristics of spiral-groove and shrouded Rayleigh step profiles for high-speed non-contacting gas seals. ASME Paper 68-LubS-38, 1968.

Clark, W. T. and **Lancaster, J. K.** Breakdown and surface fatigue of carbons during repeated sliding. 1963, *Wear*, **6**, 467–482.

Davies, A. R. and **O'Donoghue, J. P.** The lubrication of high pressure face seals. ASME Preprint 66WA.Lub-7, 1966.

Denny, D. F. Some measurements of fluid pressures between plane parallel thrust surfaces with special reference to radial-face seals. *Wear*, 1961, **4**, 64–83.

Dray, J. Inter-relationship between carbon-graphite structures and wear. *Proc. 4th NASA/Navy workshop on liquid lubricated seals*, Univ. of New Mexico, Oct. 1981.

Dziedzic, J. Siliconised graphite and silicon carbide as face materials for mechanical seals in 12 per cent boric acid nuclear service. *Trans. ASLE* 1980, **36**(11), 643–651.

Etsion, I. Dynamic analysis of non-contacting face seals. NASA Tech. Mem. 79294, 1980.

Etsion, I. and **Dan, Y.** An analysis of mechanical face seal vibrations. *Trans. ASME, J. Lubric. Tech.*, July 1981, **103**(3), 428–435.

Findlay, J. Cavitation in mechanical face seals. Master's Thesis, Union College, Schenectady, 1967.

Fisher, M. J. An analysis of the deformation of the balanced ring in high pressure radial-face seals. *1st Int. Conf. on Fluid Sealing*, Paper D4, BHRA, 1961.

Fisher, M. J. A study of the suitability of a hydrostatic seal for boiler feed pumps. BHRA Report RR776, 1963.

Fisher, M. J. and **Nau, B. S.** An experimental investigation of face seal behaviour at sealed pressures up to 3000 lb/in². BHRA Report RR878, Feb. 1967.

Flitney, R. K. and **Nau, B. S.** Seal survey: Part 1–rotary mechanical face seals. BHRA Report CR1386, Dec. 1976.

Flitney, R. K., Nau, B. S. and **Reddy, M. D.** Mechanical seal reliability study 1. BHRA Report CR2165, 1984.

Gabriel, R. P. Fundamentals of spiral groove non-contacting face seals. ASLE Paper 78-AM-3D-1, 1978.

Golubiev, A. I. and **Gordeev, V. V.** Investigation of wear of mechanical seals in liquids containing abrasive particles. *7th Int. Conf. on Fluid Sealing*, Paper B3, BHRA, 1975.

Grant, W. S. Recirculating pump seal investigation. MPR Associates Inc. Technical Report NP 351 Vol. 1, Washington, DC, 1977.

Haardt, R. Les joints d'étancheite à faces radiales. Doctoral Thesis, Univ. of Lyon, July 1975.

Hill, H. H. Post separation behaviour of mechanical systems subjected to harmonic excitation. Doctoral thesis, North Carolina State Univ., Raleigh, 1969.

Hirabayashi, H., Ishiwata, H. and **Kato, Y.** Excessive abrasion of mechanical seals caused by salt solutions. *3rd Int. Conf. on Fluid Sealing*, Paper B1, BHRA, 1967.

Hooke, C. J. and **O'Donoghue, J. P.** Elastohydrodynamic lubrication of high pressure face seals. *J. Mech. Engng Sci.*, 1968, **10**(1), 59–63.

Hooke, C. J. and **O'Donoghue, J. P.** The elastohydrodynamic lubrication of high pressure face seals. *4th Int. Conf. on Fluid Sealing*, Paper 16, BHRA, 1969.

Hughes, W. F. and **Chao, N. H.** Phase change in liquid face seals II—isothermal and adiabatic bounds with real fluids. *Trans. ASME, J. Lubric. Tech.*, July 1980, **102**(3), 350–359.

Ikeuchi, K. and **Mori, H.** Pumping action of seal-like surfaces under hydrodynamic lubrication. *J. of JSLE* (Int. Edition), Nov. 1980, **1**, 83–86.

Ikeuchi, K. and **Mori, H.** Pumping action in face seals with elastic deformation. *J. of JSLE* (Int. Edition), April 1981, **2**, 111–114.

Ikeuchi, K. and **Mori, H.** The pumping in face seals with elastic deformation–sealing of pressurised liquid surrounded with air. *J. of JSLE* (Int. Edition), 1982a, **3**, 87–90.

Ikeuchi, K. and **Mori, H.** Hydrodynamic lubrication in seals with cavitation: 1st report—effect of cavity pressure on lubricating film. *Bull. JSME*, June 1982b, **25**, 204, 1002–1007.

Ikeuchi, K. and **Mori, H.** Hydrodynamic lubrication in seals with cavitation: 2nd report—effect of cavity pressure on leakage. *Bull. JSME*, June 1982c, **25**, 204, 1008–1013.

Iny, E. H. The design of hydrodynamically lubricated seals with predictable operating characteristics. *5th Int. Conf. on Fluid Sealing*, Paper H1, BHRA, 1971a.

Iny, E. H. A theory of sealing with radial face seals. *Wear*, 1971b, **18**, 51–69.

Ishiwata, H. and **Hirabayashi, H.** Friction and sealing characteristics of mechanical seals. *1st Int. Conf. on Fluid Sealing*, Paper D5, BHRA, 1961.

Janetz, R. W. Mechanical seals in pumps for hydronic systems. *Lub. Engng.*, Sept. 1965, 372–380.

Johnson, R. L., Swikert, M. A. and **Bailey, J. M.** Wear of typical carbon-base sliding seal materials. NACA Report No.TN3595, 1956.

Kanasaki, S. and **Schoenherr, K.** Sealing boric acid solutions with mechanical seals in nuclear service. *JSLE–ASLE Int. Lubn. Conf.*, June 1975.

Kaneta, M., Fukahori, M. and **Hirano, F.** Dynamic characteristics of face seals. *8th Int. Conf. on Fluid Sealing*, Paper A2, BHRA, 1978.

Karpe, S. A. Microscopic examination of high cobalt base super-alloy face seal mating inserts. *Proc. 4th NASA/Navy workshop on liquid lubricated seals*, Univ. of New Mexico, Oct. 1981.

Kennedy, F. E. Thermocracking of a mechanical face seal. *Proc. 4th NASA/Navy workshop on liquid lubricated seals*, Univ. of New Mexico, 1981.

Kilaparti, S. R. and **Burton, R. A.** Pressure distribution for patchlike contact in seals with frictional mating, thermal expansion and wear. ASME Preprint No. 76-Lub5–11, 1976.

Kittner, C. A. and **Metcalfe, R.** An inside view of rotary seal dynamics. *Proc. 5th Symposium on Engineering Applications of Mechanics*, University of Ottawa, 1980.

Kojabashian, C. and **Richardson, H. H.** A micropad model for the hydrodynamic performance of carbon face seals. *3rd Int. Conf. on Fluid Sealing*, Paper E4, BHRA, 1967.

Kupperman, D. S. Dynamic tracking of non-contacting face seals. *Trans. ASLE*, 1978, **18**(4), 306–311.

Labus, T. J. The influence of rubbing materials and operating conditions on the power dissipated by mechanical seals. *Lubn. Engng.*, July 1981, 387–394.

Lancaster, J. K. Transition in the friction of wear of carbons and graphites sliding against themselves. *Trans. ASLE*, 1975, **18**(3), 187–201.

Lancaster, J. K. Additive effects on the friction and wear of graphite carbons. 5th Leeds–Lyon Symposium on Tribology: *The wear of non-metallic materials*, 1978, (Mech. Engng. Publns, London).

Lebeck, A. O. Causes and effects of waviness in contacting mechanical face seals. Univ. of New Mexico Tech. Report ME68(76) NSF271-1, Jan. 1976a.

Lebeck, A. O. Theory of thermoelastic instability of rotating rings in sliding contact with wear. *Trans. ASME, J. Lubric. Tech.* April 1976b, 277–284.

Lebeck, A. O. A study of mixed lubrication in contacting mechanical face seals. 4th Leeds–Lyon Symposium on Tribology: *Surface roughness effects in lubrication*, 1977, 46–57, (Mech. Engng Publns, London).

Lebeck, A. O., Teale, J. L. and **Pierce, R. E.** Hydrodynamic lubrication and wear in wavy contacting face seals. *Trans. ASME, J. Lubric. Tech.*, Jan. 1978, **100**, 81–91.

Lebeck, A. O. Hydrodynamic lubrication with wear and asperity contact in mechanical face seals. Univ. of New Mexico, Bureau of Engng. Res., Tech. Report ME95(79) ONR414-1, Jan. 1979.

Lebeck, A. O. The effect of ring deflection and heat transfer on the thermoelastic instability of rotating face seals. *Wear*, 1980a, **59**, 121–133.

Lebeck, A. O. A mixed friction hydrostatic face seal model with phase change. *Trans. ASME, J. Lubric. Tech.*, April 1980b, **102**, 133–138.

Lebeck, A. O. A mixed friction hydrostatic mechanical face seal model with thermal rotation and wear. *Trans. ASLE*, Oct. 1980c, **23**(4), 357–387.

Lebeck, A. O. Experimental evaluation of a mixed friction hydrostatic mechanical face seal model considering radial taper, thermal taper and wear. ASME Preprint No. 81-Lub-38, 1981a.

Lebeck, A. O. A test apparatus for measuring the effects of waviness in mechanical face seals. *Trans. ASLE*, 1981b, **24**(3), 371–378.

Lebeck, A. O. Hydrodynamic lubrication in wavy contacting face seals–a two dimensional model. *Trans. ASME, J. Lubric. Tech.*, Oct. 1981c, **103**, 578–586.

Lebeck, A. O. Design of an optimum moving wave and tilt mechanical face seal for liquid applications. *9th Int. Conf. on Fluid Sealing*, Paper E2, BHRA, 1981d.

Lohou, J. Hydrodynamique des joints d'étancheite du type radial. Doctoral Thesis, Univ. of Lyon, June 1970.

Lohou, J. and **Godet, M.** Angular misalignments and squeeze-film effects in radial face seals. *6th Int. Conf. on Fluid Sealing*, Paper D2, BHRA, 1973.

Ludwig, L. P. Self-acting shaft seals. *Proc. AGARD Conf. on Seal Technology in Gas Turbine Engines*, Paper 16, 1978.

Mayer, E. Unbalanced mechanical seals for liquids. *1st Int. Conf. on Fluid Sealing*, Paper E2, BHRA, 1961.

Metcalfe, R. Performance analysis of axisymmetric flat face mechanical seals. Atomic Energy of Canada Ltd, Report AECL-4432, Feb. 1973.

Metcalfe, R. End face seals in high pressure water—learning from those failures. *Lub. Engng.*, 1976a, **32**(12), 625–636.

Metcalfe, R. The use of finite element deflection analysis in performance predictions for end face seals. Atomic Energy of Canada Ltd, Report AECL 5563, July 1976b.

Metcalfe, R., Pothier, N. and **Rod, B. H.** Diametral tilt and leakage of end face seals with convergent sealing gap. *8th Int. Conf. of Fluid Sealing*, Paper A1, BHRA Fluid Engineering, 1978.

Metcalfe, R. End face seal deflection effects—the problems of two-component stationary or rotating assemblies. *Trans. ASLE*, 1980a, **23**(4), 393–400.

Metcalfe, R. Dynamic whirl in well-aligned, liquid-lubricated end face seals with hydrostatic tilt instability. ASLE Paper 80-LC-1B-1, 1980b.

Metcalfe, R. Dynamic tracking of angular misalignment in liquid lubricated end face seals. *Trans. ASLE* Oct. 1981, **24**(4), 509–516.

Metcalfe, R. *et al.* Effects of pressure and temperature changes on end-face seal performance. *Trans. ASLE*, 1982, **25**(3), 361–371.

Moores, J. and **Marsh, M. G.** The effects of eccentricity and flatness of the sealing face of a carbon mechanical contact face seal. *1st Int. Conf. on Fluid Sealing*, Paper H1, BHRA, 1961.

Nau, B. S. and **Turnbull, D. E.** Some effects of elastic deformation on the characteristics of balanced radial-face seals. *1st Int. Conf. on Fluid Sealing*, Paper D3, BHRA, 1961.

Nau, B. S. An investigation into the nature of the interface film, the pressure-generation mechanism and centripetal pumping in mechanical seals. BHRA Report RR754, 1963.

Nau, B. S. Further experimental studies of face seal hydrodynamics. BHRA Report RR855, 1966.

Nau, B. S. Observations of face seal flatness after use. BHRA Report RR877, 1967.

Nau, B. S. Centripetal flow in face seals. ASLE Paper 68-AM–4B-3 [*Lub. Engng.*, 1968, **25**(4), 161–8].

Nau, B. S. Film cavitation observations in face seals. *4th Int. Conf. on Fluid Sealing*, Paper 20, BHRA, 1969.

Nau, B. S. Observations and analysis of mechanical seal film characteristics. *Trans. ASME, J. Lubric. Tech.*, 1980, **102**(3), 341–349.

Nau, B. S. Vibration and rotary mechanical seals. *Tribology Int.*, Feb. 1981a, **14**(1), 55–59.

Nau, B. S. Vibration and rotary mechanical seals: a review. BHRA Report RR1766, Nov. 1981b.

Nau, B. S. The operation of unbalanced mechanical seals at high pressure. BHRA Report RR1819, Mar. 1982.

Nelson, W. E. Pump curves can be deceptive. NPRA Refinery and Petrochemical Plant Maintenance Conf., Farrar Assoc., Tulsa, 1980.

Netzel, J. P. Observations of thermoelastic instability in mechanical face seals. 1980a, *Wear*, **59**(1), 135–148.

Netzel, J. P. Surface disturbances in mechanical face seals from thermoelastic instability. ASLE Preprint 80AM6B-1, 1980b.

Netzel, J. P. The effect of interface cooling in controlling surface disturbances in mechanical face seals. *Wear*, 1982, **79**, 119–127.

Orcutt, F. K. An investigation of the operation and failure of mechanical face seals. *4th Int. Conf. on Fluid Sealing*, Paper 22, BHRA, 1969.

Pape, J. G. Fundamental aspects of radial-face seals. Doctoral Thesis, TH Delft, Oct. 1969.

Paxton, R. R. and **Shobert, W. R.** Testing carbon for seals and bearings. *Lub. Engng.*, Jan. 1971, 27–33.

Paxton, R. R. and **Strugala, E. W.** Wear rates of impregnated carbons sealing warm water. *Lub. Engng.*, Aug. 1973, 341–346.

Paxton, R. R., Massaro, A. J. and **Strugala, E. W.** Performance of siliconised graphite as a mating face in mechanical seals. *Lub. Engng.*, Dec. 1977, 650–656.

Paxton, R. R. and **Hulbert, H. T.** Rubbing friction in radial face seals. *Lub. Engng.*, Feb. 1980, 89–97.

Piehn, L. D. The application of tungsten carbide for mechanical face seal faces. *Lub. Engng.*, Sept. 1965, 381–385.

Pijcke, A. C. and **Vries, P.** Comparative testing of a number of radial face seals. *8th Int. Conf. on Fluid Sealing*, Paper E2, BHRA, 1978.

Rowles, R. T. and **Nau, B. S.** An assessment of factors affecting the response of mechanical seals to shaft vibration. *8th Int. Conf. on Fluid Sealing*, Paper A3, BHRA Fluid Engineering, 1978.

Ruddy, A. V., Dowson, D. and **Taylor, C. M.** The prediction of film thickness in a mechanical face seal with circumferential waviness on both the face and the seat. *J. Mech. Engng. Sci.*, 1982, **24**(1), 37–43.

Ruddy, A. V. and **Summers-Smith, J. D.** The mechanism of film generation in mechanical face seals. *Tribology Int.*, Aug. 1982, 227–231.

Sehnal, J., Sedy, J., Zobens, A. and **Etsion, I.** Performance of the coned-face end-seal with regard to energy conservation. *Trans. ASLE*, 1982, **26**(4), 415–429.

Smalley, A. J., Darlow, M. S. and **Mehta, R. K.** The dynamic characteristics of O-rings. ASME Paper 77-DET-27, 1977.

Stanghan-Batch, B. Face lubrication in mechanical seals. *Proc. 9th Tribology Conv.*, 1971, (IMechE, London).

Stanghan-Batch, B. and **Iny, E. H.** A hydrodynamic theory of radial-face mechanical seals. *J. Mech. Engng. Sci.*, 1971, **15**(1), 17–24.

Strom, T. N., Ludwig, L. P. and **Hudelson, J. C.** Vibration of shaft face seals and stabilising effect of viscous and friction damping. NASA Report TN D-5161, 1969.

Strugala, E. W. The nature and causes of seal carbon blistering. *Lub. Engng.*, Sept. 1972, 333–339.

Summers-Smith, J. D. Laboratory investigation of the performance of a radial-face seal. *1st Int. Conf. on Fluid Sealing*, Paper D1, BHRA, 1961.

Summers-Smith, J. D. Performance of mechanical seals in centrifugal process pumps. *9th Int. Conf. on Fluid Sealing*, Paper H1, BHRA, 1981.

Symons, J. D. Factors contributing to the water pump seal leakage problem. General Motors Research Laboratories Report GMR1736, Nov. 1974.

Taylor, I. and **Jones, D. N.** High duty seal tests: some observations on lubrication. *Proc. 8th BPMA Technical Conf: Pumps—the heart of the matter*, BHRA, 1983.

Teale, J. L. and **Lebeck, A. O.** An evaluation of the average flow model for surface roughness effects in lubrication. *Trans. ASME, J. Lubric. Tech.*, July 1980, **102**, 360–367.

Tribe, F. Seawater-lubricated mechanical seals and bearings: associated materials problems. *Lubn. Engng.*, 1973, **39**(5), 292–299.

Trytek, J. J. The application of mechanical end face seals for hot water service. *Lub. Engng.*, Jan. 1973, **39**(1), 17–23.

Watson, R. D. Endurance tests on large hydrostatic (conical face) and hydrodynamic (taper-face) rotating shaft seals in pressurised water. Atomic Energy of Canada Ltd, Report AECL4367, Jan. 1973.

Watson, R. D. Effect of seal ring deflection on the characteristics of face-type mechanical shaft seals in high pressure water. Atomic Energy of Canada Ltd, Report AECL2242, April 1965.

Will, T. P. Experimental observation of a face-contact mechanical shaft seal operating on water. ASLE Preprint 81-LC-1A-1, 1981.

Winn, L. W., Thorkildsen, R. L. and **Wilcock, D. F.** Design of one-piece jet-engine compressor and seal. ASME Paper 68-LubS-14, 1968.

Wu, Y. and **Burton, R. A.** Thermoelastic and dynamic phenomena in seals. ASME Preprint 80-C2-Lub-9, 1980.

Zuk, J. Fundamentals of fluid sealing. NASA Technical Note TN D-8151, Mar. 1976.

APPENDIX

Abbreviations

AECL	Atomic Energy of Canada Limited
ANSI	American National Standards Institute
ASLE	American Society of Lubrication Engineers
ASME	American Society of Mechanical Engineers
API	American Petroleum Institute
BHRA	British Hydromechanics Research Association
BS	British Standard
CEGB	Central Electricity Generating Board, UK
DIN	German industrial standard
EPRI	Electric Power Research Institute, USA
ISO	International Standards Organization
MTI	Mechanical Technology Inc, USA
NASA	National Aeronautical and Space Administration, USA
NMI	Netherlands Maritime Institute
RAE	Royal Aircraft Establishment, UK

INDEX

Page numbers in **bold** type refer to definitions of terms given in Appendix 1.